ECOLOGY WORKBOOK

Other Related Books

A Text Book of Environmental Science	*Dr. Vidya Thakur*
A Text Book on Farming Systems	*U.K. Behera*
Allelopathy in Soil Sickness	*S.S. Narwal*
Basic Ground-Water Hydrology	*U.S.D.I.*
Biofumigation and Solarization for Management of Soil-Borne Plant Pathogens	*P. Parvatha Reddy*
Challenges and Strategies of Dryland Agriculture	*Rao, S.C.*
Climate Change and its Ecological Implications for the Western Himalaya	*V.L. Chopra*
Crop Production in Waterlogged Saline Soils	*S.K. Gupta*
Indira Agriculture Competition Explorer	*R.L. Arya et al.*
New Vistas of Organic Farming	*Mukund Joshi*
Organic Agriculture in the Tropics and Subtropics	*Ulrich kopke*
Organic Agriculture	*J.C. Tarafdar et al.*
Organic Farming: Theory & Practice	*SP Palaniappan*
Plant Analysis - Research Methods	*S.S. Narwal*
Plant Analysis: Comprehensive Methods and Protocols	*B.K. Garg*
Principles in the Quantitative Analysis of Waters, Fertilizers, Plants Soils	*U.S. Sree Ramulu*
Resource Conserving Techniques in Crop Production	*A.R. Sharma*
Salinity Tolerance in Plants: Methods, Mechanisms & Management	*B.K. Garg*
Soil and Plant Analysis Laboratory Manual	*John Ryan et al.*
Soil and Plant Analysis	*C.S. Piper*
Soil and Water Conservation in Semi-arid area	*FAO*
Soil Salinity Assessment Methods & Interpretation of Electrical Conductivity Measurements	*FAO*
Soil Survey Laboratory Methods Manual	*USDA*
Soil Survey Manual - Revised and Enlarged Ed.	*USDA*
Soil-Water Engineering Field & Laboratory Manual	*Thomas J. Trout*
Standard Methods for Analysis of Soil Plant and Water	*I.C. Gupta et al.*
The Soil – Plant Systems in Relation to Inorganic Nutrition	*Maurice Fried*
Use of Saline Water in Agriculture	*I.C. Gupta*

ECOLOGY WORKBOOK

R. Misra

Professor and Head of the Department of Botany
Banaras Hindu University, Varanasi, India

SCIENTIFIC PUBLISHERS

Ecology Workbook

R. Misra

Reprint in India, 2026

by

Scientific Publishers (India)

5-A, New Pali Road, P.O. Box 91
Jodhpur - 342 001 (India)

4806/24, Ansari Road, Daryaganj
New Delhi - 110 002 (India)

www.scientificpubonline.com

ISBN: 978-81-7233-805-3

Printed in India

PREFACE

It gives me pleasure to place ECOLOGY WORKBOOK in the hands of students who have waited long for it.

The publication of 'Indian Manual of Plant Ecology' in 1955 clearly demonstrated big gaps in our knowledge for want of ecological data from the tropical countries. Since then many more students in India have shown interest in the subject. Meanwhile, the preparatory phase of the International Biological Programme has emphasized the writing of standard field and laboratory manuals.

With a view to make India's participation in the International Biological Programme effective I invited twentyfive selected University teachers in April, 1966 to join a three week 'School on Plant Ecology' held in October-November, 1966. This gave me six months' time to provide the school with the material for trial in the field and the laboratory. I looked to ten of my colleagues and students for help which was offered generously. They prepared much of the original draft manuscript, backed by their experience of teaching and research. Hence, I am thankful to all of them viz., Dr. K. C. Misra, Dr. R. S. Ambasht, Dr. S. S. Ramam, Dr. H. R. Sant, Dr. R. S. Tripathi, Dr. S. M. Mukherjee, Dr. K. P. Singh, Dr. J. S. Singh, Dr. R. R. Das and Dr. S. B. Pandey. Further, the last two persons helped me in preparing the manuscript in the final form. My wife and Sri Brij Gopal assisted me in the correction of the proof.

The 'School on Plant Ecology' was held with the financial assistance of the University Grants Commission. The participants of the school gave valuable suggestions for improving the text and I am thankful to all of them.

It is hoped that the exercises presented in the workbook will lead to better comprehension of the concept of ecosystem. They are likely to reveal the biological basis of organic productivity. With this background many of the biologists will find interest in the International Biological Programme which concerns with the biological basis of production and human welfare.

The workbook presents a series of exercises for understanding and evaluating the flow of energy and cycling of water and minerals in the three major plant ecosystems viz. grass, forest and freshwater. The exercises are given separately for each of the three ecosystems. It begins with grass since most of the University centres and laboratories have easy access to it. A practice of the exercises first in the grassland and the laboratory is likely to equip the worker with knowledge and confidence to move on to the forest or the freshwater areas for study. It has, therefore, been found convenient to use identical decimal numbering of the exercises for corresponding topics of the three ecosystems. This method has the advantage of comparing and bringing together data from the record books, for appreciating the structural and functional differences between the ecosystems. Lists of the flora of Varanasi giving habit and phenology of the species, are included in appendices A, B and C of the book corresponding to grass, forest and freshwater communities. It is hoped that these model lists will be replaced by appropriate ones for their own areas, by the workers.

I shall be grateful for suggestions for further improvement of the workbook in its future editions.

Varanasi (India) R. MISRA
July 20, 1968

CONTENTS

* Pt. I—Grassland Ecosystem
* Pt. II—Forest Ecosystem
* Pt. III—Freshwater Ecosystem

INTRODUCTION

The workbook is based on the concept of ecosystem—the structural and functional unit of ecology. A preliminary recognition of the same in the field becomes prerequisite for the following exercises.

Ecosystem: An ecosystem comprises of the interaction of organisms and their environmental factors involving flow of energy and cycling of minerals. Naturally, it has structural characteristics depending upon the spatial distribution of plants and animals. Its functional aspect includes trapping of energy by the green plants and its passage into herbivorous and carnivorous animals and other organisms which are either parasitic or saprophytic. The solar radiations originally received in the system get dissipated out of it by the respiratory activity of the participating organisms at various stages so that for maintenance of the ecosystem energy must be consistently added and transferred from organism to organism within the community. The amount of energy gained by the system and the mode and quantities of it transferred, affect the structure of the system such as density and diversity of species.

While energy flows through the ecosystem, minerals like nitrogen, phosphorus, potassium, calcium, etc., are absorbed from the soil by the plants and then passed on to the organisms in the form of elaborated compounds. The salts are released after the death and decomposition of the organisms and reabsorbed by the plants. The use and reuse of the minerals by the organisms keep them circulating within the ecosystem although appreciable amounts may be lost to the system by leaching and added by the process of weathering of the mineral particles.

The limitations imposed on the rates and amounts involved in the circulation by the climatic, edaphic and biotic factors and above all by the energy economy, react upon the structure of the ecosystem. Thus structure, function and factors of the environment become the triangular base of interaction of the ecosystem and indeed for the study of ecology. The exercises of the workbook are oriented towards evaluation of these bases in relation to plants which are the primary producers.

For practical purposes an ecosystem may be delineated in the field by taking an area of uniform topographic and vegetational pattern. An analysis of adjoining communities may indicate degrees of affinity, interdependence or divergence, justifying modifications in the limits of the hypothetical ecosystem.

Record: The following are some preliminary observations and findings which must be recorded in the beginning of the exercises:

1. Site

(*a*) Location:

Give name of the locality to be studied and find out with the aid of a standard atlas, the latitude and longitude at which it is situated. Also record the altitude by means of an altimeter or barometer. It is useful to give the mode of approach and accessibility to sites in wilderness.

(*b*) Physiography:

The physical relations of earth, air and water constitute the study of physiography in the real sense. In ecological studies, however, it includes pedology and edaphology. The former deals with the origin and the development of soil while the latter embodies the study of the soil characteristics (including topography) affecting the plant growth. Topography includes the study of surface features such as slopes, elevations, depressions, drainage lines and exposures with their directions and measurements. The extent of erosion and silting is also considered under this heading. Depth of water table and its seasonal fluctuations can be measured from nearby wells.

(*c*) Origin and development of soil:

Soil owes its origin to different types of rocks through the processes of weathering which are brought about by mechanical or chemical means. The rocks disintegrate into the primary mineral particles which yield on leaching solution system. Many of the solutes may react leading to synthesis of secondary mineral particles. With the development of life on soil, plant and animal remains are incorporated in it thus constituting the organic system of the soil. The biological system of the soil consists of living micro-organisms and larger animals and plant parts. The air spaces in the soil constitute the gaseous system. The coming together of the five systems completes the process of soil formation.

The development of soil as described above leads to its stratification or layering on account of varying forces of weathering at different depths. Soil profile is the term used to designate this layering of soil. Generally three horizons viz., A, B and C, are distinguished depending on the characters of the particles, chemical composition and colour. The uppermost layer of the soil contains leaf litter, half decomposed litter and humus. This forms the horizon 'A' which may be further subdivided into horizons Aoo, Ao, A_1 and A_2. Beneath this layer in 'B', we get an admixture of mineral particles and humus which generally impart grey to dark brown colour to this soil layer. Horizon 'C' consists of the loosened parent material and 'D' stands for the unaltered parent material which may be rock or alluvial and other deposits.

Dig one meter deep trench in the soil or remove a core of the soil by means of soil auger or borer. Mark different layers and sublayers. Take small samples from each layer and note colour, texture, structure and organic or inorganic characters of the particles. Determine the sesquioxide ratio, $SiO_2/Fe_2O_3+Al_2O_3$, and the calcium content of each layer. Draw a profile diagram indicating the determined characters.

Select varying habitats such as grassland, drying pond and forest and study the soil profiles. Compare and contrast the characters of different horizons of the selected sites and interpret the results.

(*d*) NATURE OF ROCKS:

As stated earlier, soil has been derived from rocks and hence, the study of geological rocks specially in case of residual and immature soils, seems important. The soil forming rocks may be grouped in three main categories, viz., igneous rocks, sedimentary rocks and metamorphic rocks. The description of mineralogical and chemical composition of the rocks put under the above three categories are given by Misra and Puri (1954).

(*e*) HISTORICAL:

Circumstantial evidences are sometimes conclusive in judging the status of a vegetation type. But, according to van Steenis (1961), the absence of sufficient knowledge of the history of the site is often a lamentable obstruction in this respect. Thus, the knowledge of the history of the spot or the area of study is of prime importance. The history of the site should be examined with reference to (i) anthropogenic and (ii) geological features.

(*i*) *Anthropogenic history:*

"A vegetatiographer should never underrate the influence of former occupation and destruction by man, the traces of which disappear in the tropics easily and rapidly, partly because the occupations and civilizations often used no permanent structures of stone, etc., and partly as even permanent structures fall here into decay much more rapidly......" (van Steenis, 1961). Formerly it was assumed that coastal swamps, inland country, etc., have never been inhabited but recent diggings by archaeologists have revealed opposite facts.

Record observations from:—

1. Perusal of historical records if available for any lost civilisation, land use, etc.

2. Interrogation of local people as to the previous land-use, lost village, gold-digging, abandoned fields, etc.

3. 'Scars' of human occupation, e.g., broken irrigation channels, dikes of rice fields, etc.

4. Availability of broken and old earthenwares.

5. Occurrence of fires: (*a*) existence of wood charcoal in soil, (*b*) charred tree trunks, (*c*) charred rhizomes of perennial grasses.

6. Hole digging to find out if soil profile has been disturbed in the past.

7. Abrupt changes in the vegetation: see if evidence of past fire is present on boundaries, or the soil profile is loose as due to shifting cultivation, etc.

8. Protection by man: if any species is abundant because of protection it received in the past by man.

9. Absence of thick forests which is chiefly due to deforestation by man in the past.

10. Signs of looping on cut branches.

If in the area of study the vegetation is different from surrounding areas or similar areas elsewhere, and study of anthropogenic factor has revealed interference by man, discuss the implications.

(*ii*) *Geological history*:

The study of the time factor in terms of geological history of the area is of paramount importance because it often helps in evaluating the exact relationship of vegetation with the habitat factors in time (Puri, 1960). The change of climate is also covered by such study.

For purpose of examining the geological history of the area and its relation to present day vegetation, gather the following information:—

1. Fossil records from the area: consultation of published reports and examination at the site.

2. Glaciation, time and course: published reports to be consulted.

3. Any change in the courses of nearby rivers: archaeological and geographical survey reports to be examined.

4. Sculptures depicting trees, etc., available from the area or from adjacent areas to be examined, and identification of the depicted trees to be made. This may give valuable evidence of change in climate (Randhawa 1945).

5. Past and present day distribution of dominant flora of the area.

If fossil records show occurrence of some species which is absent in the present times, the change in the climate may be discussed. Discuss differences in the vegetation between the area of study and adjacent areas, whether it could be based on geological history.

2. Classification of Vegetation:

Vegetation constitutes the total plant cover of a region, area or place. Classification of vegetation is required mainly for recording of informations in an orderly manner and for their storage, prompt recovery and discussions. Although there are many systems of classification based on the purpose, one that can be followed even with the minimum of informations is given below (Fosberg, 1961). The seven criteria generally used in classifications are: physiognomy, structure, function (features suggesting special adaptation), composition, dynamics (successional phenomenon), environmental relations (habitat), and history. Out of these structure and function have been found to be the most satisfactory criteria (Fosberg, 1961). Three major structural groups have been proposed—Closed vegetation: in at least one layer, plants predominantly touching; Open vegetation: separated by not more than twice their diameters; Sparse vegetation: more space between the plants of the most complete layer.

CLASSIFICATION

I. Major Structural Groups: Closed vegetation
(Crowns or peripheries of plants touching or overlapping)

I-A. Formation Class: Forest
(Closed woody vegatation 5 m or more tall)

I-A-1. Formation Group: Evergreen forest
(At least the canopy with no significant leafless period)

F_1 Multistratal evergreen forest (rainforest)
(Tall, multistratal, orthophyllous; epiphytes and lianas usually common; shrub layers usually sparse)
F_2 Evergreen swamp forest
(Bases of trees or root systems adapted to lengthy or permanent submergence; multistratal or unistratal; peat development often notable)
F_3 Gnarled evergreen forest
(Low, trunks and branches tending to be twisted or gnarled, usually with one tree layer, usually more or less sclerophyllous)
F_4 Evergreen hard-wood orthophyll forest
(One or two woody layers, irregular canopy, medium to low stature, many branched tree)
F_5 Evergreen soft-wood orthophyll forest
(Usually unistratal, of fast-growing trees, tangled with lianas)
F_6 Evergreen broad sclerophyll forest
(Leaves hard, stiff, or coriaceous)
F_7 Evergreen narrow sclerophyll forest (needle leafed forest)
(Leaves (or equivalent) linear or scale-like, hard)
F_8 Evergreen bamboo forest
(Dominant layer of large bamboo or giant reed)
F_9 Microphyllous evergreen forest
(Frequently but not always of compound-leafed trees, sometimes thorny)

I-A-2. Formation Group: Deciduous forest
(At least canopy layer bare of leaves for a period during cold or dry season)

F_1 Winter-deciduous orthophyll forest (Hardwood forest)
F_2 Deciduous swamp forest
F_3 Dry-season deciduous forest
(Orthophyllous; thick herb or shurb layer)
F_4 Microphyllous deciduous forest
(Trees mostly thornless or spineless)
F_5 Deciduous thorn forest
(Trees mostly armed with thorns, spines or prickles, usually microphyllous)

I-B. Formation Class: Scrub
(Closed woody vegetation 5 m or less tall)

I-B-1. Formation Group: Evergreen scrub

F_1 Evergreen orthophyll scrub
F_2 Evergreen bamboo or reed brake
(Dominant layer of dwarf bamboo, cane, or woody reeds, may be strictly erect or tangled)
F_3 Evergreen swamp scrub
(Composed of shrubs with adaptations to stand lengthy or permanent submergence of root systems; peat development may be notable)
F_4 Everrgeen broad sclerophyll scrub
(Stiff shrubs, leaves generally rather small; with abundant, epiphytes equivalent of mossy forest)
F_5 Mossy evergreen sclerophyll scrub
F_6 Gnarled evergreen narrow sclerophyll scrub (krummholz)
F_7 Straight evergreen narrow sclerophyll scrub
(Not especially gnarled)
F_8 Microphyllous evergreen scrub
(Often thorny)

I-B-2. Formation Groups: Deciduous scrub
(Shrub's periodically bare of leaves, usually in dry season or winter)

F_1 Deciduous orthophyll scrub
F_2 Deciduous swamp scrub
(Root systems adapted to prolonged submergence)
F_3 Deciduous thorn scrub (Thornbush)
(Usually microphyllous, cacti often abundant. Rare in this closed form)

I-C . Formation Class: Dwarf scrub
(Closed predominantly woody vegetation, less than 0.5 m tall)

I-C-1. Formation Group: Evergreen dwarf scrub

F_1 Evergreen orthophyll dwarf scrub
F_2 Evergreen broad sclerophyll dwarf scrub
F_3 Evergreen shrub bog
(Dwarf scrub with significant peat accumulation, root systems of plants adapted to constant immersion)

I-C-2. Formation Group: Deciduous dwarf scrub

F Deciduous orthophyll dwarf scrub

I-D. Formation Class: Open forest with closed lower layers
(Trees with crowns not touching, mostly not separated by more than twice their diameter)

I-D-1. Formation Group: Evergreen open forest with closed lower layers

F_1 Open evergreen orthophyll forest
F_2 Open evergreen swamp
(Water table at or near the ground surface)
F_3 Open evergreen broad sclerophyll forest

F_4 Open evergreen narrow sclerophyll forest
F_5 Open microphyllous evergreen forest
(May be somewhat thorny)

I-D-2. Formation Group: Open deciduous forest with closed lower layers

F_1 Open deciduous orthophyll forest
F_2 Open deciduous swamp
F_3 Open deciduous narrow sclerophyll forest
F_4 Open microphyllous deciduous forest

I-E. Formation Class: Closed scrub with scattered trees

I-E-1. Formation Group: Closed evergreen scrub with scattered trees
(At least shrub layer evergreen)

F_1 Evergreen orthophyll scrub with trees
F_2 Evergreen sclerophyll scrub with trees
F_3 Microphyllous scrub with trees

I-E-2. Formation Group: Closed deciduous scrub with scattered trees

F Deciduous orthophyll scrub with trees

I-F. Formation Class: Dwarf scrub with scattered trees

I-F-1. Formation Group: Evergreen dwarf scrub with scattered trees

F_1 Microphyllous evergreen dwarf scrub with trees
(Without significant peat formation)
F_2 Microphyllous evergreen heath with trees
(With peat accumulation)

I-F-2. Formation Group: Deciduous dwarf scrub with trees

F Deciduous heath with trees
(With significant peat accumulation)

I-G. Formation Class: Open scrub with closed ground cover

I-G-1. Formation Group: Open evergreen scrub with closed ground cover

F_1 Open evergreen orthophyll scrub
F_2 Open evergreen broad sclerophyll scrub
F_3 Open gnarled evergreen narrow sclerophyll scrub or open krummholz

I-G-2. Formation Group: Open deciduous scrub with closed ground cover

F Open deciduous orthophyll scrub with closed ground cover

I-H. Formation Class: Open dwarf scrub with closed ground cover

I-H-1. Formation Group: Open evergreen dwarf scrub with closed ground cover

F_1 Open evergreen orthophyll dwarf scrub
F_2 Open evergreen shrub bog
F_3 Open evergreen microphyllous dwarf scrub

I-H-2. Formation Group: Open deciduous dwarf scrub with closed ground cover

F_1 Open deciduous orthophyll dwarf scrub
(Without significant peat accumulation)
F_2 Open deciduous orthophyll heath
(With significant accumulation of peat)

I-I. Formation Class: Tall savanna
(Closed grass or other herbaceous vegetation 1 m tall or more with scattered trees)

I-I-1. Formation Group: Evergreen savanna
(Trees evergreen)

F_1 Evergreen orthophyll savanna
F_2 Evergreen broad sclerophyll savanna
F_3 Evergreen broad sclerophyll swamp savanna
F_4 Evergreen narrow sclerophyll savanna
F_5 Microphyllous evergreen savanna

I-I-2. Formation Group: Deciduous tall savanna
(Trees deciduous)

F_1 Deciduous orthophyll savanna
F_2 Deciduous microphyll savanna
F_3 Deciduous thorn savanna

I-J. Formation Class: Low savanna
(Herbaceous vegetation less than 1 m tall with scattered trees)

I-J-1. Formation Group: Evergreen low savanna
(Trees evergreen)

F_1 Evergreen orthophyll low savanna
F_2 Evergreen broad-sclerophyll low savanna
F_3 Evergreen narrow-sclerophyll low savanna
F_4 Evergreen narrow sclerophyll lichen savanna
F_5 Evergreen narrow sclerophyll swamp savanna

I-J-2. Formation Group: Deciduous low savanna

F_1 Deciduous orthophyll savanna
F_2 Deciduous lichen savanna
F_3 Deciduous swamp savanna

I-K. Formation Class: Shrub savanna
(Closed grass or other herbaceous vegetation with scattered shrubs)

I-K-1. Formation Group: Evergreen shrub savanna

F_1 Evergreen orthophyll shrub savanna
F_2 Evergreen broad sclerophyll shrub savanna
F_3 Evergreen narrow sclerophyll shrub savanna
F_4 Evergreen microphyll shrub savanna

I-K-2. Formation Group: Deciduous shrub savanna

F_1 Deciduous orthophyll shrub savanna
F_2 Deciduos broad sclerophyll shrub savanna
F_3 Microphyllous deciduous shrub savanna
F_4 Mesophyllous deciduous thorn shrub savanna
F_5 Microphyllous deciduous thorn shrub savanna

I-L. Formation Class: Tall grass
(Closed herbaceous vegetation exceeding 1 m in height, predominantly graminaid)

I-L-1. Formation Group: Evergreen tall grass
(Shoots remaining green the year round)

F_1 Evergreen orthophyll tall grass
F_2 Evergreen orthophyll tall tussock grass
F_3 Evergreen sclerophyll tall grass

I-L-2. Formation Group: Seasonal tall grass
(Turning brown in dry season or in winter, often burned)

F_1 Seasonal orthophyll tall grass
F_2 Seasonal sclerophyll grass

I-M. Formation Class: Short grass
(Closed herbaceous vegetation, less than 1 m tall, predominantly graminoid)

I-M-1. Formation Group: Evergreen short grass

F_1 Evergreen orthophyll short grass
F_2 Evergreen orthophyll short tussock grass
F_3 Evergreen short grass and sedge bog
(Growing from a mass of grass and sedge-peat)

I-M-2. Formation Group: Seasonal short grass

F_1 Seasonal orthophyll meadows (short grass)
F_2 Seasonal sclerophyll short grass meadow

I-N. Formation Class: Broad-leafed herb vegetation
(Closed vegetation, predominantly of broad-leafed herbaceous plants)

I-N-1. Formation Group: Evergreen broad-leafed herb vegetation

F_1 Evergreen broad-leafed weedy vegetation
F_2 Evergreen fern meadow
F_3 Evergreen giant herb thicket

I-N-2. Formation Group: Seasonal broad-leafed herb vegetation

F Seasonal broad-leafed meadow

I-O. Formation Class: Marsh
(Closed vegetation of herbaceous plants, the roots of which are adapted to lengthy or permanent immersion, the upper parts to emersion.)

I-O-1. Formation Group: Evergreen marsh

F_1 Evergreen orthophyll graminoid marsh
F_2 Tall evergreen graminoid marsh
F_3 Evergreen broad-leafed marsh

I-O-2. Formation Group: Seasonal marsh

F_1 Seasonal orthophyll marsh
F_2 Seasonal sclerophyll marsh

I-P. Formation Class: Submerged meadows
(Vegetation of rooted aquatic herbs, adapted for permanent complete submersion except in some cases for floating leaves)

I-P-1. Formation Group: Evergreen submerged meadows

F_1 Evergreen watergrass
F_2 Macrophyllous evergreen submerged meadows
F_3 Megaphyllous evergreen submerged meadows

I-P-2. Formation Group: Seasonal submerged meadows
(Plants disappearing, at least their shoots, in winter)

F_1 Seasonal watergrass
F_2 Broad-leafed seasonal submerged meadows
(Leaves of ordinary size, not narrowly linear or grasslike)
F_3 Macrophyllous seasonal submerged meadows
(Water lilies and similar plants dominant)

I-Q. Formation Class: Floating meadows
(Closed vegetation of aquatic herbs, adapted to floating conditions, not rooted in bottom)

I-Q-1. Formation Group: Evergreen floating meadows

F_1 Broad-leafed evergreen floating meadows
F_2 Evergreen floating grass
F_3 Thalliform evergreen floating vegetation
F_4 Microphyllous evergreen floating meadows

I-Q-2. Formation Group: Seasonal floating meadow

F_1 Thaliform seasonal floating meadow
F_2 Microphyllous seasonal floating meadow

II. Primary structural Group: Open Vegetation
(Plants or tufts of plants not touching but crowns not separated by more than twice their diameter; plants, not substratum, dominating landscape)

II-A. Formation Class: Steppe forest (often called woodland or woodland-savanna)
(Tree layer and lower layers open, lower layers may be open or sparse)

II-A-1. Formation Group: Evergreen steppe forest
(Tree layers, at least, evergreen)

F_1 Evergreen broad sclerophyll steppe forest
F_2 Open evergreen sclerophyll swamp
F_3 Evergreen narrow sclerophyll steppe forest
F_4 Microphyllous evergreen steppe forest

II-A-2. Formation Group: Deciduous steppe forest

F_1 Deciduous orthophyll steppe forest
F_2 Microphyllous deciduous steppe forest
F_3 Deciduous bamboo steppe forest
F_4 Open deciduous thorn (steppe) forest

II-B. Formation Class: Steppe scrub
[Like steppe forest, but with shrubs (over 0.5 m tall) instead of trees]

II-B-1. Formation Group: Evergreen steppe scrub

F_1 Evergreen broad sclerophyll steppe scrub
F_2 Gnarled evergreen narrow sclerophyll steppe scrub
F_3 Microphyllous evergreen steppe scrub
F_4 Evergreen succulent steppe scrub
F_5 Evergreen saltbush steppe scrub
(Shrubs with gray scurfy leaves)

II-B-2. Formation Group: Deciduous steppe scrub

F_1 Deciduous orthophyll steppe scrub
F_2 Microphyllous deciduous steppe scrub
F_3 Deciduous thorn steppe scrub (thornbush)

II-C. Formation Class: Dwarf steppe scrub
(Open predominantly woody vegetation less than 0.5 m tall)

II-C-1. Formation Group: Evergreen dwarf steppe scrub

F_1 Evergreen orthophyll dwarf steppe scrub
F_2 Evergreen broad sclerophyll dwarf steppe scrub
F_3 Evergreen narrow sclerophyll dwarf steppe scrub
F_4 Microphyllous evergreen dwarf steppe scrub

II-D. Formation Class: Steppe savanna
(Steppe with scattered trees)

II-D-1 Formation Group: Evergreen steppe savanna
(Trees evergreen)

F_1 Evergreen orthophyll steppe savanna
F_2 Evergreen sclerophyll steppe savanna

II-D-2. Formation Group: Deciduous steppe savanna

F_1 Deciduous orthophyll steppe savanna
F_2 Microphyllous deciduous steppe savanna
F_3 Deciduous thorn steppe savanna

II-E. Formation Class: Shrub steppe savanna

II-E-1. Formation Group: Evergreen shrub steppe savanna

F_1 Evergreen orthophyll shrub steppe savanna
F_2 Evergreen sclerophyll shrub steppe savanna
F_3 Evergreen microphyll shrub savanna
F_4 Evergreen succulent shrub steppe savanna

II-E-2. Formation Group: Deciduous shrub steppe savanna

F_1 Deciduous orthophyll shrub steppe savanna
F_2 Microphyllous deciduous shrub steppe savanna
F_3 Deciduous thorn shrub steppe savanna

II-F. Formation Class: Dwarf shrub steppe savanna

II-F-1. Formation Group: Evergreen dwarf shrub steppe savanna

F_1 Evergreen narrow sclerophyll dwarf shrub steppe savanna
F_2 Succulent dwarf shrub steppe savanna

II-F-2. Formation Group: Seasonal dwarf shrub steppe savanna

F Seasonal sclerophyll dwarf shrub steppe savanna

II-G. Formation Class: Steppe
(Open herbaceous vegetation, tufts or plants discrete, yet close enough to dominate the landscape)

II-G-1. Formation Group: Evergreen steppe

F_1 Evergreen saltbush steppe
F_2 Evergreen succulent steppe
F_3 Evergreen cushion plant steppe

II-G-2. Formation Group: Seasonal steppe

F_1 Seasonal grass steppe
F_2 Annual herb steppe

III. Primary Structural Group: Sparse vegetation or desert
(Plants so scattered that substratum dominates landscape)

III-A. Formation Class: Desert forest
(Scattered trees, subordinate shrub or herb layers very sparse, or absent)

III-A-1. Formation Group: Evergreen desert forest
(May be evergreen because of persistent leaves or because of green stems)

F_1 Evergreen non-succulent desert forest
F_2 Evergreen succulent desert forest

III-A-2. Formation Group: Deciduous desert forest

F_1 Microphyllous deciduous desert forest

III-B. Formation Class: Desert scrub
(Scattered shrubs in an otherwise bare or only ephemerally vegetated landscape, not here differentiated into shrub and dwarf shrub classes)

III-B-1· Formation Group: Evergreen desert scrub

F_1 Evergreen sclerophyll desert scrub
F_2 Microphyllous evergreen desert scrub
F_3 Saltbush desert
F_4 Evergreen desert thorn-scrub
(Dominated by rigid green-stemmed, spinose leafless, or ephemerally leafy plants)
F_5 Evergreen succulent desert scrub

III-B-2. Formation Group: Deciduous desert scrub

F_1 Microphyllous deciduous desert scrub
F_2 Deciduous desert thorn scrub

III-C. Formation Class: Desert herb vegetation
(Scattered herbaceous plants only)

III-C-1. Formation Group: Evergreen desert herb vegetation

F_1 Evergreen succulent herb desert
F_2 Evergreen psammophyte desert
(Plants with special adaptations enabling them to survive in shifting sand)
F_3 Lichen tundra

III-C-2. Formation Group: Seasonal desert herb vegetation

F_1 Seasonal desert grass
F_2 Ephemeral herb desert
(Vegetation principally of ephemeral annuals and geophytes, appearing only for short periods after infrequent rains)

3. Requirements

The requirements of field and laboratory are listed in the Appendix. In addition to these it is necessary to be equipped with suitable note books, proforma, charts, graph papers, maps, etc. These may be conveniently filed within aluminium covers which have the advantage of protecting the papers from rain and can also serve as a hard back for writing. The file may be suspended by the neck with a pencil tied to it.

One should not forget to carry a first-aid medical box with the party. A vasculum and plant press are necessary adjuncts. The vasculum is also useful for making noise to drive away wild animals and collect the party men together when they have separated in a forest. An inflatable boat and rubber suiting or gum boots are useful for work on lakes.

Instructions for laboratory work are given in the manuals of laboratory plant physiology. One such manual is 'Plant in action' by Machlis and Torrey (1956).

4. Preparation of report

Reports should invariably be written on the completion of an exercise, Honesty and accuracy in presentation of data always pay dividends. Reasoning and self-thinking are the main traits of a scientist. A general plan for presentation of report would include (a) title of the exercise, (b) dates of commencement and completion, (c) objectives, (d) procedure, (e) observations and (f) discussion. The discussion examines the achievements of the objectives in light of the observations and is the soul of any exercise.

REFERENCES

Ashby, M. 1963. *Introduction to Plant Ecology*. Macmillan & Co. Ltd., New York.

Fosberg, F. R. 1961. A Classification of vegetation for general purposes. *Trop. Ecol.* **2**: 1-28.

Gates, F. C. 1949. *Field Manual of Plant Ecology*. McGraw Hill Book Co. Inc., New York.

Machlis, L. and Torrey, J. G. 1956. *Plants in Action*. W.H. Freeman & Co., San Francisco.

Misra, R. and Puri, G. S. 1954. *Indian Manual of Plant Ecology*. Oxford Book Co., New Delhi.

Puri, G. S. 1960. *Indian Forest Ecology*, Vol. I. Oxford Book Co., New Delhi.

Randhawa, M. S. 1945. Progressive desiccation of northern India in historical times. *Bomb. Nat. Hist. Soc.* **45**: 558-568.

van Steenis, C. G.G. J. 1961. Axioms and criteria of vegetatiology with special reference to the tropics. *Trop. Ecol.* **2**: 33-47.

1. GRASSLAND ECOSYSTEM

1.00 HABITAT STUDY

1.01 Mapping of the Site

It is sometimes desirable to relocate accurately the area which has been studied in the past. This could easily be done with a map made at that time. Base maps are also necessary for putting data on the occurrence and distribution of plants/ vegetation types. A number of methods for mapping are available. Simpler ones are given below:—

Procedure A: Select a point of vantage from which various parts of the area under study are observable. Take a graph sheet and put various boundaries of vegetation types, depressions, elevations, etc. on a rough scale, and label them properly. Drainage pattern may also be shown. Give emphasis to permanent objects like outcrops and large trees, etc.

Such sketch maps are most useful for small areas because detailed notes of things observed in the field can be made on them at the time of study.

Procedure B: Traverse the area in several directions and put in stakes or flags at a number of points. Make sure that one stake is visible from the other. Now measure the distances between the stakes, 1-2, 2-3, 1-3, 2-4, etc. Tabulate the data as below:—

Points	Distance (m)	Remarks

Continue the measurements till there are at least two measurements from each point and the end point coincides with the starting one.

When mapping a lake or a large pond, put the stakes around the same.

Now take a graph sheet and mark points (stakes) on them using the measured distances. Within this mapped area smaller stakes may be used to demarcate the boundaries of vegetation types or depressions, etc. Record separate measurements for these and fill up accordingly on the base map prepared earlier.

Procedure C: Set up a base line of convenient length on one side of the area to be mapped and measure it. Put in one stake at each end of the line and place compasses on tripods over there. Adjust the compasses so that the bearings are reciprocal with each other. Record these bearings.

Select a number of points (representing features like path, road, pool, ridge, particular tree, or vegetation boundary, etc. and label them in succession. With the help of an assistant who takes a rod to the first point read and record the bearings in the two compasses. Repeat the procedure for second and successive points in order. Record the data in the following table:—

Points	Compass reading from one end	Compass reading from the other end

Now prepare a sketch map of the area noting the location of points and other interesting features (Exercise 1.01/A).

Prepare the map with the help of two protractors placed at the two ends of the base line drawn to scale on a paper. The protractors are so adjusted as to designate the recorded bearings of the respective compasses. Draw lines from each end of the base line and put point A where the two lines cross each other. Mark other points in order. Complete the details using the sketch map.

This method may also be used successfully for mapping ponds and lakes. The points are then placed at several places across the water border.

Precautions: Each point (stake) should be visible from both the ends of the base line. If it is not so then subsidiary line (s) will have to be used.

Procedure D (Traverse method): Put in a stake at the starting point and place the compass mounted on a tripod over it. Select any point in the area and instruct the assistant to move there with a stadia rod. Sight this point (with the help of the stadia rod there) through compass and record the bearing. Measure distance between the compass and the point taken. Now move the compass to this point and select another point in the area for placing a stadia rod. Record the bearing and distance again. Repeat this procedure till the area has been reasonably covered. When mapping a lake or pond the starting point should coincide accurately with the end point. Tabulate data as below:—

Points	Bearing	Distance	Remarks

Prepare the map with the help of the data recorded in the above table.

If need be, permanent objects like large trees, steep elevation, etc., may be taken as points. This will exclude the use of stadistet rods.

Procedure E (Contour mapping): Set up a line at right angles to a section line (or other line agreed upon) of the area. Run a number of traverses at measured intervals and record the data as earlier (Ex. 1.01/D). Along these traverses, at measured distances, elevation is recorded by means of an altimeter.

Lay out the base map as in the traverse method and put in the figures of elevation at particular points. Thereafter smooth out contours. Check this map in the field.

Various vegetation types can be conveniently shown on this map.

Precaution: Contour lines should never cross each other.

Procedure F (Aerial photographs): Select a few important landmarks in the area which can be recognised in the aerial photographs and measure distances between them. When the aerial photograph is available, prepare the map with the help of these landmarks. The landmarks will help in accurate scaling of the map.

1.02 Physical Characters of Soil

A. **Temperature:** Soil temperature is measured by using soil thermometer. The bulb is buried to the desired depth and the temperature can be read on the dial.

B. **Soil Colour:** Soil colour frequently regarded as an index of soil fertility has been the basis of naming big soil groups. It also helps in distinguishing soil horizons. Soil colour is mainly due to the presence of iron and manganese and organic matter.

The Munsell soil colour chart contains standard colour chips arranged systematically on cards carried in a loose note book. Three simple variables, namely hue, value and chroma combine to give almost all the colours. Hue of a colour indicates its relation to spectral colours like red, yellow, green, etc., value refers to its relative lightness, and the chroma gives its strength (or departure from the neutral of the same lightness).

Spread the soil uniformly over a cardboard and select the card which contains matching colour chips. Each card carries colour chips of constant hue, designated by the symbol (letter abbrevation of the spectral colour) in the upper right hand corner of the card. Each vertical row shows colours of increasing value. In horizontal rows colours increase in chroma towards right. The numbers printed below each colour chip indicate its value and chroma respectively. Locate the nearest matching colour chip. In case of variegated or mottled soil the dominant colour should be indicated. Moisten the soil and determine the changed colour. In writing the Munsell notation, order is hue, value, chroma with a space between the hue letter and the succeeding value number, and a virgule between the two numbers for value and chroma. Record the colour name and Munsell notation for dry and moist soils as given below:

Reddish brown (5YR 4/4; 3/4, moist)

Brown (10YR 5/3; 4/3, moist).

REFERENCE

Soil Survey Manual, U.S. Dept. Agriculture Handbook No. 18, 1951.

C. Soil Texture: Soil texture refers to the relative proportion of the various groups of individual particles. Particles more than 0.02 mm in diameter, viz. sand and gravel mainly provide physical support to the plants. Amount of silt and clay particles (less than 0.02 mm dia.) determine the water holding capacity and nutrient availability of the soil. Soils are classified into a number of textural classes depending upon their particle size distribution. Texture can be evaluated approximately in the field or accurately in the laboratory by mechanical analysis. For details of mechanical analysis see Piper (1944).

International System

Textural group	Size in mm
Coarse Sand	2.00—0.20
Fine Sand	0.20—0.02
Silt	0.02—0.002
Clay	below 0.002

Soil sample is examined under hand lens and felt between fingers in dry as well as moist state.

Examine the soil under hand lens for large particles. Squeeze small amount of air dry soil as well as moist soil between thumb and finger and try to judge the texture with the help of key given below:

Sand: Sand is loose and coarse grained. The individual grains can readily be seen or felt. Squeezed when moist, it will form a cast, but will crumble when touched.

Sandy Loam: A sandy loam is a soil containing much sand but which has enough silt and clay to make it somewhat coherent. The individual sand grains can readily be seen and felt. Squeezed when dry, it will form a cast which will readily fall apart, but if squeezed when moist a cast can be formed that will bear careful handling without breaking.

Loam: A loam is a soil having a relatively even mixture of different grades of sand, silt and clay. It is mellow with a somewhat gritty feel, yet fairly smooth and slightly plastic. Squeezed when dry, it will form a cast that bear careful handling, while the cast formed by squeezing the moist soil can be handled quite freely without breaking.

Silt Loam: A silt loam is a soil having a moderate amount of the fine grades of sand and only a small amount of clay, over half of the particles being of the size called 'silt'. When dry it may appear cloddy but the lumps can be readily broken, and when pulverized it feels soft and floury. When wet the soil readily runs together and puddles. Either dry and moist it will form casts that can be freely handled without breaking, but when moistened and squeezed between thumb and finger it will not 'ribbon' but will give a broken appearance.

Clay Loam: A clay loam is a fine textured soil which usually breaks into clods or lumps that are hard when dry. When the moist soil is pinched between the thumb and finger it will form a thin 'ribbon' which will break readily, barely sustaining its own weight. The moist soil is plastic and will form a cast that will bear much handling. When kneaded in the hand it does not crumble readily but tends to work into a heavy compact mass.

Clay: A clay is a fine textured soil that forms very hard lumps or clods when dry and is quite plastic and usually sticky when wet. When the moist soil is pinched out between the thumb and fingers it will form a long, flexible 'ribbon'. Some fine clays very high in colloids, are friable and lack plasticity in all conditions of moisture.

REFERENCES

Piper C. S. 1950: *Soil and Plant Analysis, Adelaide.*

Soil Survey Manual, U.S. Dept. Agriculture, Handbook No. 18, 1951.

D. Pore Space: Interstitial space around the soil particles is known as pore space. Part of the pore space is filled with water, while the rest is occupied by air. The amount of pore space greatly depends upon the texture, compaction and aggregation of soil particles. Light textured soils usually contain less pore space than heavy textured soils.

The difference in true specific gravity and apparent specific gravity (volume weight) forms the basis of estimation of pore space.

Scrape the soil surface to a flat plain. Dig a straight walled pit (e.g. $10 \times 10 \times 10$ cm) and collect the soil removed in a beaker. This soil sample is oven dried at 105°C and weighed. Find out the volume of the pit as follows. Fill the measuring cylinder with sand dropped through a funnel. Then fill the pit completely by dropping sand from the measuring cylinder through the same funnel. Note the volume of sand needed.

CALCULATION

Calculate percentage pore space by the formula given below:

$$\text{Per cent pore space } \frac{2.6-D}{2.6} \times 100$$

where D is the volume weight of soil and is calculated as oven dry weight of soil sample removed from the pit divided by the volume of the pit, and 2.6 is the approximate specific gravity of soil particles.

Note: Instead of digging a pit, soil of known volume can be sampled by employing core sampling cylinders. Wilde, Voigt and Iyer (1964) have described several types of core samplers.

REFERENCES

Daubenmire, R.F. 1947. *Plants and Environment.* John Wiley, New York.

Wilde S. A., G. K. Voigt and J. G. Iyer, 1964. *Soil and Plant Analysis for Tree culture.* Oxford Publishing House, Calcutta.

1.03 Chemical Characteristics of Soil

I. RAPID TESTS FOR SOILS

Procedure A: Water Content is determined by drying the sample in an oven at 105°—110°C. In our climate, it is a deciding factor in the distribution of species, especially during the dry seasons, when in some cases it may fall to 0.5% of the dry weight. During the rainy season, water content of the soil is chiefly a function of the topography of the land. But in the dry seasons, it depends more on clay fraction of the soil as organic matter is too low.

Procedure B: Organic matter is variously determined but none of the methods is entirely satisfactory. The determination of nitrogen is best done by Kjeldahl's process in which the SO_2 fumes may be absorbed by iodine for carbon estimation of the same sample. Our soils are precariously poor in their organic content. Carbon content of most of the samples is only 0.6% and nitrogen is rarely more than 0.06%. On account of the low organic content the soils are incapable of supporting a rich and healthy population of bacteria. The same factor is responsible for poor water content of more open soils in the dry seasons. Natural peat formation is found no where. The biotic factors are mainly responsible for removal of organic matter from the soil surface; whatever remains behind, is rapidly decomposed as the process is favoured by a high temperature and bright light. The white ants and micro-organisms do their job.

Procedure C: Carbonate Content is roughly determined in the field by noting the degree of effervescence produced by adding HCl to the sample. It is done accurately in the laboratory by Collin's calcimeter. The saline soils are rich in NO_3 and CO_3. Very few species like *Butea frondosa*, *Aristida* spp. and a few other grasses can grow on such Usar lands. Soil with calcareous matter as present in 'kankar', is also common. Many plants growing on these show chlorosis; and where a hard pan is formed underneath, it is difficult to establish tree seedlings.

Procedure D: Nitrate Content is examined in the field on a clear 1: 5 soil: water solution with diphenylamine in conc. H_2SO_4 (0.2%) by adding 2 drops of each on a white spot plate. The depth of the blue colour produced is arbitrarily noted in 4 degrees. Accurate estimations in the laboratory can be done by the phenoldisulphonic acid method.It should be noted that nitrate production in the soil may be fairly rapid, yet negative results may at times be obtained if the sample is collected soon after rains, when it is washed out of the soil, or when it is collected under a dense growth where it is quickly absorbed by the plants. Hence, as suggested by Olsen, the samples after collection should be kept for 24 hours under nearly natural conditions before the analysis. Nitrate production is very rapid when muds which remained under water for a long time get exposed on the drying of a pool. Soils rich in organic matter are generally rich in nitrates also. *Rumex dentatus*, *Trianthema monogyna*, *Portulaca oleracea*, *Lippia nodiflora*, etc. are definitely nitrophilous species. Absorption of nitrate by a plant is dependent on water content of drier soils. Thus *Euphorbia hirta* on suffering from low nitrate absorption becomes red in the dry seasons.

Procedure E: Deficiency of exchangeable bases like Na, K, Ca and Mg is easily determined in the field by the modified Comber's test. It consists of shaking vigorously a pinch of soil with a saturated alcoholic solution of ammonium thiocyanate in a test tube, and keeping it aside for the solids to settle down, when a drop of H_2O_2 is added to the clear supernatant liquid. Development of red colour in the solution and its depth, as empirically noted in 4 degrees, give the deficiency of the replaceable bases. Waterlogged soils show this deficiency and it is probable that aquatic and marsh vegetation rooted to the bed of a pond are adjusted to such conditions.

Procedure F: Reductivity or reducing capacity of waterlogged samples is reckoned by noting the difference in the colours produced before and after adding H_2O_2 in the above mentioned test. It depends on the fact that such samples are mostly base deficient and a good proportion of the exchangeable iron is present in the reduced ferrous state which, on getting oxidised with H_2O_2, intensifies the colour due to the formation of ferric thiocynate. It is an important property of the waterlogged soils and, as many of our soils become so in the rainy season, the test is generally recommended. An extremely reducing soil will compete for oxygen against the growing roots. However, Conway finds that oxygen gas diffuses from the leaves down to the roots in many of the marsh plants, particularly the sedges and the grasses. But liberation of aluminium in the exchange complex due to effects of leaching, proves highly toxic to plant growth. Nitrate production altogether stops in such soils, and poisonous gases like H_2S and CH_4 are generally formed here, if there is enough of organic matter to decompose.

The same property of the soil has been more accurately determined by finding the redox potential and calculating the results in rH. values. These values range from 0—41 with a balance between oxidation and reduction at 20.5. Values above this are found for oxidising soils and lower values stand for reducing samples. The extreme values for subaquous muds and oxidised agricultural soils are nearly 7 and 31 respectively. It was thought at one time that oxidising values are related to soil fertility but experiments at Rothamsted could not substantiate it. Actually agriculturists had never experimented with the reducing soils. For, it was found that many of the extremely reducing muds in the English Lakes were quite fertile for a large number of aquatic species and these, when exposed to dryness, obtained highly oxidising values. On the other hand, sterile peats had values near about 18 which, on oxidation, increased only up to 22 or so, showing their inertness. Thus it is clear that rapid and steep drifts in rH values on oxidation or reduction are more correlated with fertility, rather than the actual rH values. A definite relationship between rH of soil and plant distribution and plant succession has already been established for aquatic plants in the English Lakes.

II. SOIL pH

pH is the most important single property identifying the chemical nature of soil.

The pH of the soil depends upon the relative amounts of adsorbed hydrogen and adsorbed metallic cations. When hydrogen ions predominate the reaction becomes acidic. If reverse is true alkalinity results. Soil pH is correlated with amount of calcium and magnesium, solubility of iron, aluminium, manganese and phosphorus, and activity of micro-organisms.

Procedure A (Universal Indicator Method): Universal indicator is a mixture of different indicators with having different critical pH. It shows whole range of spectrum colours from red to violet for pH values ranging between 3 and 11 respectively.

Shake thoroughly a small spoonful of the soil sample, equal amount of barium sulphate and about 20 ml of water in a test tube. Allow to stand. Place a few drops of the clear supernatant liquid on a porcelain tile and add equal amount of Universal indicator. Match the colour of the solution with the colour chart for different pH provided on the indicator bottle.

Procedure B (Electrometric Method): The electrical potential developed across a glass membrane, when a glass electrode is immersed in the soil suspension, is measured. This potential is proportional to the difference in pH between both sides of the membrane. Usually the instrument is calibrated to read directly in pH units.

REAGENTS

Standard buffer solution of pH 4.00 (0.05 M potassium biphthalate). Dissolve 10.2 g reagent grade potassium biphthalate in water and dilute to 1 litre.

Weigh out 20 g of soil sample and transfer to a 125 ml wide mouth bottle. Add 100 ml of water, fix the stopper and shake mechanically for 1 hour.

Carefully read the working instructions of the pH meter. Standardize the pH meter with the buffer solution. Remove electrodes from the buffer solution and rinse with water. Shake the bottle and dip the electrodes into the soil suspension. Read the pH.

PRECAUTIONS

Glass electrodes are broken or spoiled if handled without care. Wash the electrodes with water immediately after use and during storage keep them immersed in distilled water.

III. ELECTRICAL CONDUCTIVITY OF THE SOIL SOLUTION

The conductance of the soil solution gives a clear idea of the total soluble salts of the soil. The soil solution is prepared by shaking 5 g of soil in 100 ml distilled water. The conductivity meter is connected to 220 volt A.C. main. The first toggle switch is now turned to on and the second to 50 cycles. The two terminals marked 'cell' are connected to the terminals of the conductivity cell. This cell is immersed in the soil solution so that the platinum foils are completely under water. Set the multiplier switch at an intermediate position and rotate the main dial control the full length of the scale. Observe the shadow angle in the magic eye. If the shadow is not obtained set the multiplier switch to another position and obtain the maximum shadow angle by rotating the main dial control. The conductance of the solution is obtained by multiplying the main dial reading by the value of the multiplier switch position.

A calibration curve can be drawn by taking a number of solutions of known concentration of potassium chloride. From this curve the total soluble salts in terms of KCl can be read out against the conductance of the soil solution.

REFERENCE

Willard, H. H., L. L. Merritt, & J. A. Dean, 1958. *Instrumental methods of analysis.* D. Van Nostrand Co., N.Y.

IV. BASE EXCHANGE CAPACITY AND EXCHANGEABLE BASES

Procedure A: Ammonium saturated soil is leached with potassium chloride. Ammonium received in the leachate is the measure of cation exchange capacity.

Wash the soil leached with N ammonium acetate (as described in exercise 1.72 for preparation of soil extract), with five 50 ml portions of 95% ethyl alcohol. Then leach the soil with 450 ml of 10% acidified potassium chloride solution and collect the leachate in a 500 ml volumetric flask. Make to volume with potassium chloride and determine ammonium in a suitable aliquot (20—50 ml) either colorimetrically (see exercise 1.61/B or by distillation, (see exercise 1.61/A).

CALCULATIONS

Calculate milli equivalent of ammonium per 100 g of soil, which will represent its cation exchange capacity.

REFERENCE

Jackson, M. L. 1962. *Soil Chemical Analysis.* Asia Publishing House, Bombay.

Procedure B: (Total Exchangeable Cation—Bray and Willhites Method): Exchangeable cations are removed from the soil by leaching with ammonium acetate. After evaporation and ignition of ammonium acetate, extracted cations are converted to their carbonates or oxides, which are determined by dissolving them in standard hydrochloric acid and titrating the excess of acid.

Transfer 50 ml aliquot of ammonium acetate extract (see Exercise 1.72) into a 100 ml silica basin and evaporate to dryness on a waterbath. Ignite in a muffle furnace at 550—600°C. Cool and completely dissolve the residue in excess of 0.05 N hydrochloric acid (20—50 ml) with slight heating. Titrate the excess of acid with 0.05 N sodium hydroxide using methyl red as indicator.

CALCULATION

Calculate the amount of total exchangeable cations in soil by the formula given below:

$$\text{Exchangeable cation per 100 g soil} = (Va - Vb) \times N \times \frac{500}{V} \times \frac{100}{W}$$

where Va = ml, standard hydrochloric acid taken
Vb = ml, standard sodium hydroxide used in titration
N = Normality factor
Ve = ml, extract used
W = g, weight of soil extracted.

REFERENCE

Piper, C. S. 1950. *Soil and Plant Analysis.* Adelaide.

1.04 Atmosphere

I. TEMPERATURE

Temperature is generally measured by a mercury thermometer. Various types of direct and remote reading temperature recorders and thermograph for automatic recording are commercially available. Maximum: minimum thermometers are reset preferably at midnight or 4 to 5 P.M. each day.

II. RELATIVE HUMIDITY

The relative humidity of air is the ratio of the quantity of vapour present in any volume of the air to the quantity which would have been present in the same volume had the air been completely saturated. It is also the ratio of the elastic force of vapour at the temperature of the surrounding air. To avoid fractions, it is convenient to multiply this ratio by 100.

Instruments used for the purpose of measuring the quantity of aqueous vapour in the atmosphere are called hygrometers.

In a simple hygrometer, the instrument, consists of two thermometers as nearly as possible identical, the one marked dry and the other wet. The bulb of the wet thermometer is covered with thin muslin, and a conducting thread of lamp wick, of common darning cotton or floss silk is twisted loosely or tied in a loose knot round the neck of the bulb and over the muslin. This passes to an adjacent vessel of water placed at such a distance as to allow a length of conducting thread of about seven centimeters. The reservoir of water should be placed on one side of and a little beneath the wet bulb so that evaporation from water may not affect the reading of the dry bulb by its close vicinity.

Those instruments in which a long cistern of glass, occupying the central space between the two thermometers, is used as a reservoir for the supply of water, are undesirable since the water in the cistern becomes heated or cooled above or below the surrounding temperature and accordingly vitiates more or less the readings of the thermometers on either side.

INSTALLATION AND OBSERVATION

1. The instrument should be mounted in an open space, with the bulbs raised about 1.3 meter above the soil, in shade, and as far from walls, trees etc., as possible.

2. The water reservoir should always be supplied with rain or distilled water.

3. If the temperature of the air lowers below 32°F (O°C), the wet bulb thermometer will for a time read higher than the dry bulb. Reject such observations.

Let the water surrounding the wet bulb freeze. Then it will give the correct reading.

4. The conduction wick should be washed in a solution of sodium carbonate before use.

5. While observing the eve should be placed on a level with the top of the mercury and the observer should refrain from breathing at that time.

CALCULATIONS

From the records of Dry and Wet bulb temperatures, the relative humidity can be known by consulting hydrometrical tables.

CONVERSION OF READINGS OF WET AND DRY BULB THERMOMETER INTO RELATIVE HUMIDITIES

Temp. °C.	Depression of wet bulb thermometer in degrees Centigrade									
	1.0	2.0	3.0	4.0	5.0	6.0	7.0	8.0	9.0	10.0
0	81	64	46	29	13					
+3	84	69	54	40	25	12				
+6	87	73	60	47	35	23	11			
+9	88	76	65	53	42	32	22	12		
+12	89	78	68	58	48	38	30	21	12	4
+15	90	80	71	62	53	44	36	28	20	12
+18	90	82	73	65	57	49	42	35	27	20
+21	91	83	75	67	60	53	46	39	32	26
+24	92	85	77	70	63	56	49	43	37	31
+27	93	86	79	72	65	59	53	47	41	36
+30	93	86	79	73	67	61	55	50	44	39
+33	93	86	80	74	69	63	57	52	47	42
+36	93	87	81	75	70	64	59	54	50	45
+39	94	87	82	76	71	65	60	56	52	47
+42	94	88	82	77	72	67	62	58	53	49
+45	94	88	83	78	73	68	64	60	54	51
+48	94	89	84	78	74	69	65	61	56	53

Different types of direct reading and automatic recording hygrometers are commercially available, i.e. hair hygrometer, paper strip hygrometer, hygrograph etc.

III. FLOW, VELOCITY AND DIRECTION OF WIND

Procedure A: To measure total air movement at a place, the tolalising cup anemometer is used. It can record the length of air column passing through the instrument in a given period. By mechanical and electrical arrangements it can also show the velocity of air at a given time.

Procedure B:—To measure the velocity of unidirectional wind, fan wheel anemometer is used. It directly reads the speed as mile or kilometer per hour or per minute.

Procedure C: To record the direction of wind, wind vanes are installed which can directly show the direction at the time of observation. In fields the direction being known with the help of a magnetic compass, the wind flow can be known by leaf flutter, bending of the plant or by the movement of light plant parts, a piece of paper, smoke etc.

IV. PRESSURE OF THE AIR

The air pressure can be measured by reading the height of the mercury barometer or from the dial of Aneroid barometer. Automatic recording barograph is commercially available.

1.05 Biotic Factors

The change of human habit from food-collection to food-production has brought about many changes in the biosphere. The predominant use of the non renewable resources by human beings has complicated the matter further more. A perusal of the human history reveals that his agricultural activities have not only replaced the vegetation but have also changed the drainage system. These changes have caused natural imbalances which had gone out of human control. All the great civilizations of the past with huge populations are stretching in vast deserts in the form of ruins. This clearly signifies that in old days those areas must have been fertile enough to feed the population of the place. The human influence has caused so much of imbalance that those civilizations have been lost.

In the present day, progress of modern civilization depends more on the surplus production of the land. This has not only to sustain the populations dependent upon agriculture but also our vast and complex society of industrial workers, servicemen, professional 'hangers-on' and their domesticated animals. Therefore, it seems important to look into the ways in which the lands in the study area are being utilized and to know the total human and animal population which depends upon the land under study.

LAND USE

The records are maintained in Tahsils and from there the land use map of the desired area can be prepared.

HUMAN AND ANIMAL POPULATION

The census offices maintain a clear record of all the areas and from there the human population of the area under study can be known.

The animal populations can be ascertained as has been indicated in other parts of the book under each ecosystem. The population of domesticated animals has, however, to be collected by door to door inquiry.

INTER-ECOSYSTEM LINKS

It has already been indicated that in the present day a small area of land has to sustain a large population. The human population freely draws upon the resources of all ecosystems. There are many animals who by their movements keep on transferring energy and material resources from one ecosystem to the other.

Procedure A (Loss of timber and other produce from forest):

Nowdays all the records of the transportation of timber etc. are maintained by the forest department and the data can be collected from there.

The grazing by the domesticated animals is another source by which the produce of the forests may enter into another ecosystem. This could be estimated by knowing the populations of domesticated animals in the neighbouring villages.

Procedure B (Loss of the produce from aquatic ecosystem):

The estimates of fish catches and *Trapa* production must be made.

The transportation of plants and animals from the aquatic ecosystem by birds is also important.

The sewage disposals in the rivers are transported from one ecosystem to the other along with the streams. Thus the produce of one ecosystem can be carried over to long distances and deposited there.

Procedure C: Examine animal droppings, animal bodies and intestines of the migratory birds to show the linking of one ecosystem with the other.

Procedure D: Grazing animals leave their droppings, which may affect in two ways—firstly by increasing the nutritional status of the soil, and secondly by bringing the disseminules of new species which pass through the alimentary canal of animal undamaged.

Small mounds of dung with vigorous growth of plants on them are found. Compare the specific composition of these mounds with that of the adjacent area. Also note the differences in the growth performance of the individuals on the two types of substrata.

Plant propagules are of importance. Man has been one of the most efficient means in linking the ecosystems.

1.10 COMMUNITY STRUCTURE

Plants growing together have mutual relationships among themselves and with the environment. Such a group of plants in one area forms a *stand.* Several similar stands represent a *community.* Community is a part of an ecological system (the ecosystem) in which transformation, accumulation, and flow of energy are involved. The functioning of this system is intimately related with the components of the community. The components vary in quality as well as in quantity and impart a *structure* to the community.

The structure of a community can be studied by taking into consideration a number of characters which are usually grouped under two heads, viz., analytic and synthetic. Certain analytical characters, viz., frequency, density, abundance and dominance can be expressed quantitatively; while others, viz., sociability, vitality, periodicity and stratification find only qualitative expression. Synthetic characters include presence, constance and fidelity of components and may be computed from analytical characters of several stands of a community.

The analytical characters of a community are determined by means of three main sampling units—area, line and point, as employed in quadrat, transect and point methods respectively.

I. QUADRAT METHOD

Quadrat is the name given to the sampling unit, an area of a definite size. In shape it may be circular, rectangular or square. Depending upon the objective, the quadrats can be of many kinds, viz., list, count, chart, clip and permanent. The first three types are mainly used in the study of community structure.

Since the assemblage of plants in a community is largely heterogeneous, the determination of size and number of quadrats necessary for adequate sampling, is a pre-requisite.

Procedure A (Size of Quadrat) : With a piece of string and 3 nails form a L-shaped structure in the field. Thereafter using another piece of string and a nail differentiate an area of 10×10 sq. cm. Note down the species present within this area. Now enlarge the area, to 20×20 sq. cm and note down the additional species coming in. Again increase the area to 30×30 sq. cm and record new species. Thus go on increasing the area, and recording the additional species till you make up a 2×2 sq. m large quadrat. Summarize your data in a table.

Area	*Total no. of species*
10×10 sq. cm	
20×20 ,,	
30×30 ,,	
200×200 ,,	

Now on a graph paper plot the number of species on the Y-axis against the areas on the X-axis. The point at which the curve starts flattening up, represents the minimal area of the quadrat required for sampling that field. The curve known as "species area curve" is illustrated in figure 1·10/A.

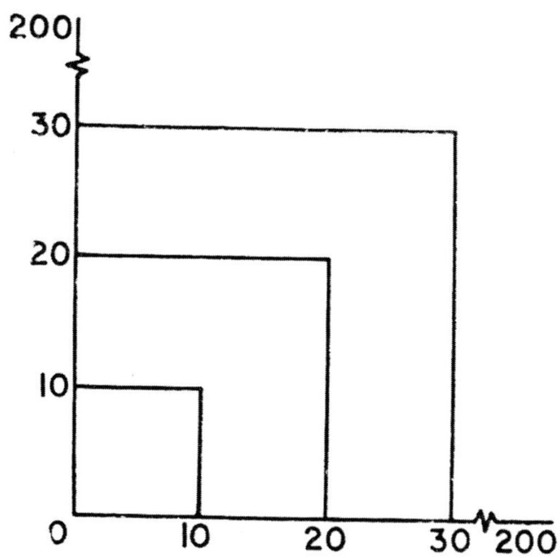

FIG. 1·10/Aa—Determination of requisite size of quadrat

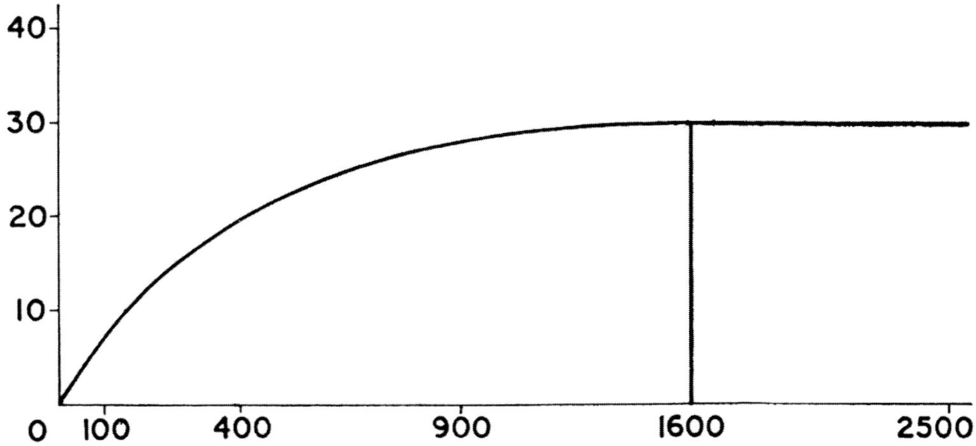

FIG. 1·10/Ab—Size of quadrat in sq. cm Species-Area Curve

REFERENCE

Oosting, H. J. 1958. *The Study of Plant Communities.* W. H. Freeman & Co., San Francisco.

Procedure B (Number of Quadrats): When the requisite size of the quadrat has been ascertained, lay 30 to 50 quadrats of that size in a stand under study. Record the species occurring in each quadrat. Tabulate your data as in procedure A, utilising the first column for number of quadrats.

Plot the number of quadrats on the X-axis of a graph sheet against the number of species on Y-axis. Here also the point at which the curve starts flattening up, gives the minimum number of quadrats required for adequate sampling of all the stands of the same community.

REFERENCE

Oosting, H. J. 1958. *The Study of Plant Communities.* W. H. Freeman & Co., San Francisco.

II. TRANSECT METHOD

A transect is a sampling strip extending across a stand or several stands (Oosting 1958). It is useful where one is concerned with the analysis of vegetation changing in composition through an ecotone (transitional zone between any two communities) or along the gradient of some environmental factor. Depending upon the objective one may employ "line transect" (also known as "line intercept") or "belt transect". The size and number of transects vary with the extent of the area to be studied.

REFERENCE

Oosting, H. J. 1958. *The Study of Plant Communities.* W. H. Freeman & Co., San Francisco.

Procedure A (Line Transect): A measuring tape or calibrated string is run across the vegetation to be analysed and the plants touching and along the tape are recorded as also the length of the tape intercepted by the individual plant cover. From this basic data various analytical characters are computed from calculations as are dealt with in subsequent excercises. This method is useful especially for bunch or tussock grasses. The construction of a profile or bisect diagram is also possible along a transect.

Procedure B (Belt Transect): In this case instead of the line, a belt of definite width (usually one meter) is studied across the vegetation. This strip is divided into a number of small sections and plant species and number of individuals of each species coming in each section are recorded. In simple words it is laying of quadrats along a line. By mapping the sections on graph paper a useful picture of the vegetation is obtained.

III. POINT METHOD

The sampling unit in this method is represented by a point. On this basis two

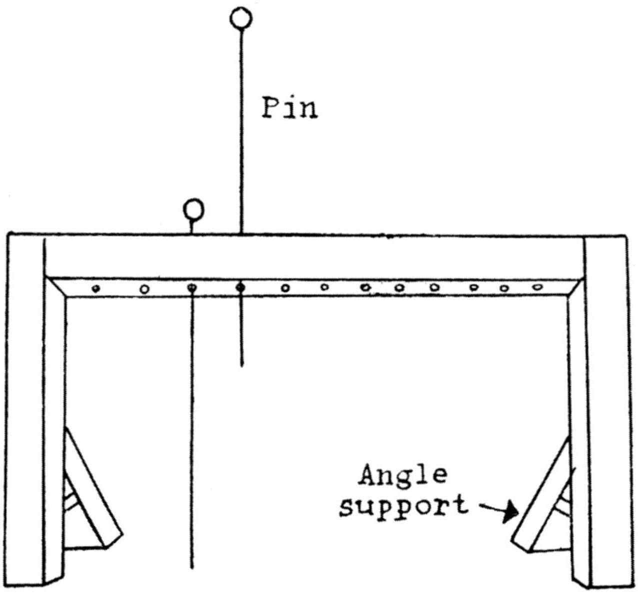

The Point-frame

FIG. 1·10/III A

main methods for the vegetational analysis have been developed: (1) Point-frame method and (2) Point-centered-quarter method.

Procedure A (Point-Frame Method): A wooden frame in which 10 movable pins are inserted at 45° angle is constructed. The usual size of a pin is 50 cm and the length of the frame is also the same. The height at which this frame is fixed on two folding wooden-crosses is also 50 cm.

The apparatus (known as point-frame) is placed at a number of places in the field, usually at random. Each time the plants hit by one or more of the ten pins, when they are lowered, are recorded. From this information various analytical characters are computed by calculations.

This method is useful for extensive grassland surveys.

REFERENCE

Phillips, E. A. 1959. *Methods of Vegetation Study*. Henry Holt & Co. Inc.

Procedure B (Point-Centered-Quarter Method): In this method of hits as in the case of the previous one, the distance of the nearest plant in each quarter from the base of the pin (needle) and its diameter are recorded. The needle is not mounted on any frame. Here, the method as modified by Misra (1964) is described.

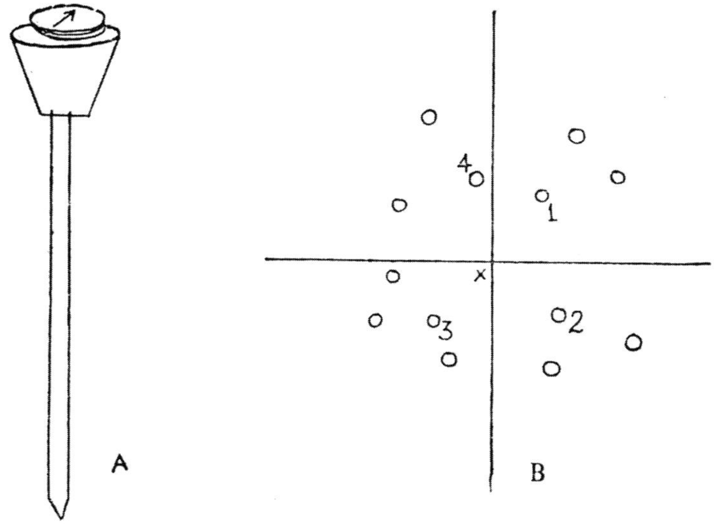

FIG. 1·10/III B

(A) Surveyor's needle.

(B) Position of surveyor's needle at X and the four quarters are indicated by imaginary lines intersecting at X, in each of the four quarters the plants nearest to the needle are indicated by 1 to 4.

A surveyor's needle (a brass rod of 3 mm thickness and 30 cm length) is provided with a compass at the top and is fixed by the worker at a point in the grass field. Around this point the area is visually divided into four quarters. In each quarter the distance between the needle and the nearest occurring plant and its diameter are recorded on the data sheet. In case the needle directly hits some plant, the distances in all the four quarters are recorded as zero and only the diameter of that particular plant is recorded. For adequate sampling the number of such points should be as high as to make provision for the measurement of at least 30 individuals of a particular plant species (Cottam *et al.* 1953, 56). The points are usually distributed at random. The method is especially suited for study of bunched grasses.

REFERENCES

Cottam, G. and J. T. Curtis. 1956. The use of distance measures in phytosociological sampling. *Ecology* **37** : 151-60.

——, ———and B. K. Hale. 1953. Some sampling characteristics of population of random dispersed individuals. *Ecology* **34** : 741-57.

Misra, K. C. 1964. Quantitative estimation of grassland vegetation through Point-Centered-Quarter method. *Trop. Ecol.* **5** : 10-16.

1.11 Frequency

Frequency as introduced by Raunkiaer (1934) indicates the number of sampling units in which a given species occurs; and thus expresses the distribution or dispersion of various species in a community. From this, percentage frequency is calculated as follows.

$$\%\ \text{frequency} = \frac{\text{Number of sampling units in which the species occurred}}{\text{Total number of units studied}} \times 100$$

Procedure A (Quadrat Method): Lay the required number of quadrats (Exercise 1.10/IB) of appropriate size (Exercise 1.10/IA) in the grassland to be studied. Note presence or absence of each species by + or – respectively in each of the quadrats.

(1) S.No.	(2) Plant Species	(3) Quadrats 1 2 3 4 5 6 7 8 9 10	(4) Total no. of quadrats of occurrence	(5) Total no. of quadrats studied	(6) Percentage frequency	(7) Frequency class

CALCULATION

Divide the value in column 4 of the above table with that in column 5 and multiply by 100 to obtain percentage frequency of each species.

From the above data prepare a frequency diagram for the field under study as described below:

Divide the species into 5 classes based on percentage frequency (Raunkiaer 1934)

Class A 1— 20%
B 21— 40%
C 41— 60%
D 61— 80%
E 81—100%

Take a graph sheet and represent the percentage of the total number of species coming into each class on Y axis against the five frequency classes on X axis, as shown in figure.

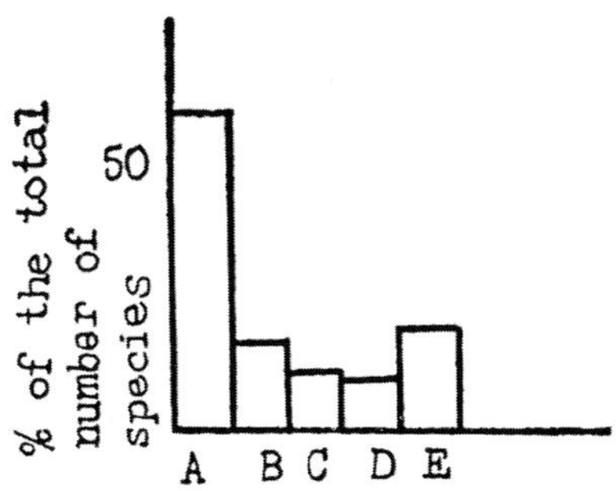

FIG. 1.11/A—Raunkiaer's normal frequency diagram.

INTERPRETATION

Raunkiaer's law of frequency class is $A > B > C \gtreqless D < E$. Frequency diagram represents the homogeneity or heterogeniety of a community as floristic uniformity varies directly with the value for class A and E. When classes B, C and D are relatively high, the stand is heterogeneous. In general, higher the class E greater will be the homogeneity.

(*i*) Name two most frequent species.

(*ii*) Whether the grassland under study is homogeneous or heterogeneous?

Procedure B (Line Transect): Run the measuring tape across the grassland in North-South direction. Record the plant species touching the tape in alternate one meter segments. Lay a number of such transects.

Record the presence or absence of each species by + or — respectively in each segment.

(1) S.No.	(2) Plant Species	(3) Segments 1 2 3 4 5 6 7 8 9 10	(4) Total no. of segments of occurrence	(5) Total no. of segments studied	(6) Percentage frequency	(7) Frequency class

CALCULATION

$$\text{Percentage frequency} = \frac{\text{Value in column 4}}{\text{Value in column 5}} \times 100$$

Divide the species in frequency classes and prepare frequency diagram as in the quadrat method.

Procedure C (Belt Transect): In this method, instead of recording the plant species which touch the tape, the plants present in alternate belt segments are listed and the percentage frequency is calculated as in procedure B.

Procedure D (Point-Frame Method): Place the point-frame apparatus at twenty places (if the area is very large increase the number) and note down the species hit by the needles each time. Also record the number of hits secured by each species. Now record the data in the following table.

(1) S. No.	(2) Plant Species / Place numbers	(3) Number of hits per place 1 2 3 4 5 6 7 8 9 10	(4) Total no. of hits secured	(5) Total no. of hits	(6) Percentage frequency	(7) Frequency class

Calculate the percentage frequency as follows for each species

$$\frac{\text{Total no. of hits secured by the species}}{\text{Total no. of hits taken}} \times 100$$

Treat the data for preparation of frequency diagram as given in procedure A.

REFERENCE

Raunkiaer, C. 1934. *The Life Forms of Plants and Statistical Plant Geography:* being the collected papers of C. Raunkiaer. Clarendon Press, Oxford, England.

1.12 Density

The terms abundance and density represent the numerical strength of species in the community. The former, if considered alongwith frequency, gives an idea of the distribution pattern of the species while the latter represents the number of individuals per unit area. The density and frequency taken together are of prime importance in determining community structure and have a variety of uses far beyond those of other quantitative values (Oosting 1958).

Procedure A (Quadrat Method): Lay a number of quadrats of desired size and count the individuals of each species in each quadrat. Record the data in the following manner.

Size of quadrat:

(1) Plant species	(2) No. of individuals quadrat number 1 2 3 4 5 6 7 8 9 10	(3) Total no. of individuals	(4) No. of quadrats of occurrence	(5) Total no. of quadrats studied	(6) Abundance	(7) Density

Now calculate abundance for each species by the following formula.

$$\text{Abundance} = \frac{\text{Value in the column 3}}{\text{Value in the column 4}}$$

Calculate density as follows:

$$\text{Density} = \frac{\text{Value in the column 3}}{\text{Value in the column 5}}$$

(Abundance is described as the number of individuals per quadrat of occurrence and density as the number of individuals per quadrat).

Specify the area of the quadrat at the top of the table and express the density as number of individuals per square meter.

Procedure B (Belt Transect): Lay transects of suitable length and width in the grassland and count the number of individuals of each species in every alternate segment of 1 meter length. Treat each segment as a quadrat. Record the data and calculate abundance and density as in procedure **A**.

Procedure C (Indirect Method): In the lower frequency ranges (up to about 12%) there is a linear relationship between frequency and density (Brown 1954). Blackman (1935) reported that for randomly distributed species the density of a species can be calculated from its percentage absence (100-percentage frequency) since there is a linear relationship between the logarithm of percentage absence and density. Kylin (1926, quoted by Blackman 1935) used the following expression.

$$e^{-m} = {}^{-ky}e$$

where m is the mean number of individuals of the species in the sampling unit, k is the density and y is the size of quadrat. The relation between the percentage absence and density is given as below

$$\text{Percentage absence} = e^{-m}$$

Thus if the quadrat size and percentage absence are known, density can be calculated. The method gives accurate measure when quadrat size is such that 80% frequency value is obtained (Blackman 1935).

REFERENCES

Blackman, G. E. 1935. A study of statistical methods of the distribution of species in grassland associations. *Ann. Bot.* **49**: 749-74.

Brown, D. 1954. *Methods of Surveying and Measuring Vegetation.* Comm. Agric. Bur. England.

Kylin. 1926. In Blackman (1935).

Oosting, H. J. 1958. *The Study of Plant Communities.* W. H. Freeman & Co., San Francisco.

1.13 Cover and Basal Area

I. COVER

The term cover or herbage cover signifies primarily the area of ground occupied by the above ground parts of plants, such as leaves, stems, and inflorescence, as viewed from above (Hanson and Churchill 1961). Cover is often the most suitable expression for recording change and is also a good measure for the availability of herbage. Usual methods for its estimation are (1) through Chart quadrat, (2) Line intercept, and (3) Point—frame methods.

Procedure A (Chart Quadrat): Divide the frame (quadrat) into 10×10 sq. cm blocks with the pressed cord. If required, as in the case of very dense and heterogeneous stands, use smaller divisions. Outline the exact areas occupied by individual plants within the quadrat, on a suitably scaled graph sheet showing the exact position of individuals. Use some identifying symbol for each species. Now total up the area covered by individuals of different species to obtain area covered by each species. The area may be measured either by counting the divisions on the graph sheet or preferably through a planimeter if available. Express the result as sq. cm per sq. m.

If a pantograph is available, set it on a low table beside the quadrat and chart the exact area through this instrument on a graph sheet or plain paper and make calculations as in the previous case.

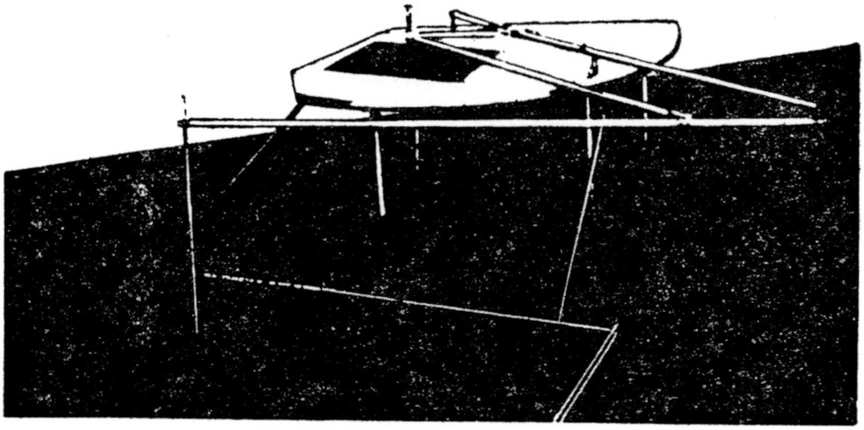

The improved pantograph set up as for use in the field.
(*Pearse, Pechanec and Pickford,* 1935)

FIG. 1·13/A

REFERENCE

Pearse, K., J. S. Pechanec and G. D. Pickford, 1935. An improved pantograph for mapping vegetation. *Ecology* **16** : 529—530.

Procedure B (Line Intercept Method): Run the measuring tape (line transect) in the grass field. Measure the length of this tape intercepted by individual plants. Total up the individual values to obtain the total length covered by different species and separately calculate the percentage of total length of the transect covered by and each species. The values obtained are percentage cover.

Procedure C (Point-Frame Method): Place the apparatus at a number of places in the field (see Excercise 1.10/III B) and note down the number of hits obtained by each species in the like manner. Use the following expression (Levy and Maiden 1933) to obtain the percentage cover of each species if the frame has been placed at 10 places.

$$\frac{(\text{No. of times a species is hit per 100}) \times (\text{100 bare spaces})}{\text{Total No. of times vegetation is hit per 100}}$$

REFERENCES

Hanson, H. C. and E. D. Churchill, 1961. *The Plant Community*. Reinhold Publishing Corporation, New York.

Levy, E. B. and E. A. Maiden, 1933. The point method of pasture analysis. *N. Z. J. Agric.* **49** : 267-79.

Oosting, H. J. 1958. *The Study of Plant Communities*. W. H. Freeman & Co., San Francisco.

Tuiney, F. W., O. S. Admodt and H. L. Ahlgren, 1937. Preliminary report of a study on methods used in botanical analysis of pasture swards. *J. Amer. Soc. Agron.* **29** : 835-40.

II. BASAL AREA

Basal area refers to the ground actually penetrated by the stems, and is readily seen when the leaves and stems are clipped at the ground surface (Hanson and Churchill 1961). It is one of the chief characters determining dominance and the nature of the community. This is measured either at 2.5 cm above ground or actually on the ground level. The values obtained in latter case are less flexible. Most common methods for determination of basal area are (1) Measurement through calipers, (2) Line-interception and (3) Point-centered quarter method.

Procedure A. (Actual Measurement): Select 100 stems of each species cut at the ground surface by sharp scissors and measure the diameters of their cut ends. Calculate average diameter for each species. Convert the values for diameter (cm) to basal area (sq. cm) through the following relation:

$$\text{Average basal area} = \pi r^2, \text{ where r (radius)} = \frac{\text{average diameter}}{2}$$

By using quadrat determine density for each species and multiply this value with that of the respective basal area. Express the result as sq. cm per sq. meter.

Procedure B (Line-Intercept Method) : Follow the same procedure as described for measurement of herbage cover (Excercise 1.13/IB), but in this case measure the length intercepted by stem only just above the ground. Make similar calculations for percentage cover.

Procedure C (Point-Centered Quarter Method): Refer exercise 1.14/B. Prepare a list of species in the decreasing order of their basal cover and herbage cover, Discuss the finding and compare with the data from other sites.

REFERENCE

Hanson, H. C. and E. D. Churchill, 1961. *The Plant Community*. Reinhold Publishing Corporation, New York.

1.14 Importance Value Index and other Characteristics

In order to express the dominance and ecological success of any species, with a single value, the concept of importance value index has been developed, This index utilises three characterstics, viz., relative frequency, relative density, and relative dominance.

Procedure A (Importance Value Index) Evaluate frequency, density and basal area for all the species growing at the study site through methods already described and calculate the following value (Phillips 1959).

1. Relative dominance $= \dfrac{\text{Total basal area of the species}}{\text{Total basal area of all species}} \times 100$

2. Relative density $= \dfrac{\text{Number of individuals of the species}}{\text{Number of individuals of all species}} \times 100$

3. Relative frequency $= \dfrac{\text{Number of occurrences of the species}}{\text{Number of occurrences of all species}} \times 100$

Now add the above three values to get importance value index (IVI). Calculate the index for all the species and arrange them in decreasing order of values to bottom. Compare the values of IVI of important species from different sites, differing in the intensity of biotic disturbance or amount of moisture in soil and discuss.

REFERENCE

Phillips, E. A. 1959. *Methods of Vegetation Study*. Henry Holt & Co. Inc.

Procedure B (Analytical Characters): One of the distance measurement methods evolved by Cottam and Curtis (1956) utilised to calculate the relative frequency and density, total density of the vegetation, absolute density, total density of the individual species, basal area and importance value index, through various expressions, as described by Misra (1964), is outlined here.

Place the needle at any spot in the field. Divide the area around the needle into four quarters. Within each quarter record the distance and diameter of the closest living plant from the needle in cm on the proforma given below. In case the needle base touches a plant, record the diameter of that plant and record the distance in all the four quarters as zero. Repeat the process at a number of spots in the field.

Compile the field data on the distance and diameter measurements of the plants on a separate proforma and calculate the following values by respective expressions.

$$\text{Relative density} = \frac{\text{Number of plants of the species}}{\text{Total No. of plants of all species}} \times 100$$

$$\text{Relative frequence} = \frac{\text{Number of points of occurrence of the species}}{\text{Total No. of points used}} \times 100$$

$$\text{Mean distance between the plants (cm)} = \frac{\text{Sum of distance of all species}}{4 \times \text{number of points used}}$$

Mean area occupied by a plant = $(\text{Mean distance})^2$ =sq. cm per plant

$$\text{Total density of plants per sq. m} = \frac{\text{1 sq. m}}{\text{Mean area}}$$

= 10,000, sq. cm Mean area—Plants per sq. m

Absolute density of individual species =

$$\frac{\text{Relative density of the species}}{100} \times \text{total density of all species.}$$

Average diameter of one plant of a species =

$$\frac{\text{Sum of diameter of all plants of the species}}{\text{Total no. of plants of that species}}$$

$$\text{Average radius of plant} = \frac{\text{Average diameter}}{2}$$

Average basal area of the plant $= \pi r^2$ (sq. cm)

PROFORMA FOR FIELD DATA

Number of points

Species	1	2	3	4	5	6	7	8	9	10	Total distance
1											
2											
3											
4											
5											
6											
7											
8											
9											

Sum of distances=

PROFORMA FOR COMPUTED DATA

Plant species	Number of points	Number of plants	Absolute density	Average diameter per plant	Area per plant	Relative frequency	Relative density	Relative basal cover	I V I
1									
2									
3									
4									
5									
6									
7									
8									
9									
10									

Total basal area (in sq. cm) of the plant species per sq. meter $= \pi r^2 \times$ absolute density

Relative basal cover of individual species $=$

$$\frac{\text{Basal cover of individual species}}{\text{Total basal cover of all the species}} \times 100$$

Importance value index = Relative density + Relative frequency + Relative basal cover

INTERPRETATION

Use this method on a number of sites and note the differences in the analytical characteristics of principal species on different sites.

REFERENCES

Cottam, G. and J. T. Curtis, 1956. The use of distance measures in phytosociological sampling. *Ecology* **37**: 451-460.

Misra, K. C. 1964. Quantitative estimation of grassland vegetation through Point-centered Quarter method. *Trop. Ecol.* **5**: 10-16.

Procedure C (Contagious distribution): Fracker and Brischle (1944) state that basic population of an area may be considered to have reached the area by random distribution of seed, and that therefore, the population at a given time consists of two components, the original plants, and offsprings of those plants. Plant populations show over dispersion either by vegetative propagation or slow rate of dispersal around the parent plants. Cole (1946) advocates the use of the terms contagious, and negatively contagious rather than over and under dispersion. Ashby (1948) states that recognition of over dispersion as a common phenomenon is a noteworthy advance, and the study of it is of major importance. Whitford (1948) used the ratio of abundance and frequency as a measure of contagion among plant populations.

$$F = \text{Frequency} = \frac{\text{No. of occupied quadrats}}{\text{Total number of quadrats}}$$

$$D = \text{Density} = \frac{\text{Total number of individuals}}{\text{Total number of quadrats}}$$

$$A = \text{Abundance} = \frac{\text{Total number of individuals}}{\text{Number of occupied quadrats}}$$

$$A \times F = 100 \times D$$

High frequency $\times$ low abundance = regular distribution

Low frequency $\times$ High abundance = contagious distribution

So the ratio of abundance to frequency is a relative measure of degree of contagiousness of the distribution of any species.

FIELD DATA FORM

S. No.	Plant species	Q_1		Q_2		Q_3		Q_4		Q_n	
		F	D	F	D	F	D	F	D	F	D

F. Note down presence or absence of plant species by + or — notation.

D. Record the number of individuals of a species in each quadrat.

Individuals = all aerial stems in most plants. In case of sedges and grasses, count each tuft as an individual.

LABORATORY DATA FORM

S. No.	Plant species	F	D	A	A/F	Contagious or regular distribution

NOTES: (*a*) Choose regularised sampling procedure.

(*b*) Set up 4 to 5 transects on the study area in a compass direction avoiding margins, depressions and depositions.

(*c*) Lay out quadrats (50×50 cm) at equidistant intervals along the transect.

(*d*) Collect at least data for 30 quadrats in any sampling site.

(*e*) Process the data for F, D, and A of individual species.

(*f*) Find out A/F ratio for every species.

(*g*) State clearly whether a species shows contagious or regular distribution.

REFERENCES

Ashby, E. 1948. Statistical Ecology. II. A reassessment. *Bot. Rev.* **14:** 222-234.

Cole, La M. C. 1946. A theory for analysing contagiously distributed population *Ecology* **27:** 329-341.

Fracker, S. B. and H. A. Brischle. 1944. Measuring the local distribution of Ribes. *Ecology* **25:** 283-303.

Procedure D (Presence): Presence is a measure of regularity of distribution of a species in different stands of an association (Phillips 1959). When list of species for a number of stands is available, presence is calculated in the manner as frequency from quadrats.

Divide the number of stands in which a given species occurs by the total number of stands and multiply by 100. Repeat this for all the species and express the results in the following scale (Braun-Blanquet 1932).

1. 1 – 20 percent
2. 21 – 40 ,,
3. 41 – 60 ,,
4. 61 – 80 ,,
5. 81 – 100 ,,

INTERPRETATION

Explain the significance of species of high and low presence.

Procedure E. If all the stands being considered have been sampled with plots of the same size, then, when the number of stands containing a species is compared to the total number of stands, the resulting quantity is known as constancy (Phillips, 1959). If there is a combination of several species with a high constancy percentage the stands belong to one association.

INTERPRETATION

Discuss whether stands under study belong to the same association.

REFERENCES

Braun-Blanquet, J. 1932. *Plant Sociology* (Trans., rev., and ed. by G. D. Fuller and H. S. Conard). Macmillan Book Co. Inc., New York.

Du Reitz, G. 1930. Vegetation furschung soziationsanalytischer Grundlage. *Handle der biol. arbeitg-meth.* teil 5, heft 2. Abderagalden.

Phillips, E.A. 1959. *Methods of Vegetation Study*. Henry Holt & Co., Inc.

Procedure F (Fidelity): Fidelity is defined as exclusiveness or the degree to which species are confined to or do better in one or a few associations. This concept is used to determine the characteristic species of an association. The scale as used by Zurich-Montpellier school is given below:

A. Characteristic species.

Fid.5 Exclusive species : species limited almost completely to one community.

Fid.4 Selective species: species found most frequently in a certain community but also, though rarely, in other communities.

Fid.3 Preferential species: species present in several communities more or less abundantly but predominantly or with better vitality in one certain community.

B. Companions.

Fid.2 Indifferent species: species without pronounced affinities for any community.

C. Accidentals.

Fid.1 Strange species: species that are rare and accidental intruders from another plant community or relics of a preceding community.

Assign fidelity numbers to various species on the basis of presence, abundance and vitality.

INTERPRETATION

Describe each association on the basis of high fidelity.

REFERENCE

Phillips, E. A. 1959. *Methods of Vegetation Study.* Henry Holt & Co., Inc.

1.15 Comparison of Stands, Ordination, Continuum, etc.

Sometimes it becomes necessary to determine whether the stands studied belong to the same community or how far two different communities resemble each other. This is done through the concept of community coefficient. In this concept any quantitative phytosociologic character of all the species present in the two stands is taken up and an index is calculated.

Procedure A (Comparison of two communities): Determine frequency of all the species growing in the two stands/communities and tabulate the data as below:

FREQUENCY PERCENTAGE OF SPECIES PRESENT IN BOTH THE STANDS/COMMUNITIES

Species	Stand A	Stand B
1		
2		
3		
4		
etc.		
Total	a	b

Now underline the values for the species which are common to both the stands. Total up the lower of each pair (A and B) of frequency percentage of these common species and designate the sum as 'W'. Total up all the values under columns A and B separately and designate the sum as 'a' and 'b' respectively. Now calculate the community coefficient by the following formula (Kulczynski 1937).

$$\text{Community coefficient } \frac{2W}{a+b} \times 100$$

Higher the coefficient more similar will be the stands to each other.

If more than two stands are to be compared, proceed taking two stands at a time.

INTERPRETATION

Study two stands of the same community and two of different communities. Explain the differences in the values of community coefficient in the two sets.

REFERENCES

Kulczynski, S. 1937. In Oosting, H. J. 1958. *The Study of Plant Communities*, W. H. Freeman and Co. San Francisco.

Jaccard, P. 1912. The distributions of the flora in the Alpine zone. *New Phytol.* **11:** 37-50.

Procedure B (Ordination): When a large number of plant communities/stands of any ecosystem have been examined, any existing pattern depending upon the behaviour of individual species may be brought about through the technique of ordination. This pattern may later be compared to any environmental factor.

Calculate the percentage of the maximum value (for the character chosen) for each species and convert these into relative values (so that the total for all the species in each stand comes to 100) for different stands separately. Arrange these values ('scores') against respective species under various columns representing different stands in a table. Now compare the scores for each species in one stand with those in another and record the lower of the two values. Thus a series of numbers will become available. Sum of these numbers (ranging from 0 – 100) will express as to how similar the two stands are with each other. Higher the value closer will be the similarity. Enter these values in the following matrix and complete it by comparing each stand with the other.

	STAND NUMBER									
	1	2	3	4	5	6	7	8	etc.	Total
1	+									—
2		+								—
3			+							—
4				+						—
5					+					—
6						+				—
7							+			—
8								+		—

Add up the rows individually and choose the stand representing lowest of these values (most unlike stand as compared to others) as stand 'a'. Examine all the values in the matrix for this stand and select as stand 'b' that stand which represents lowest value. Stand 'a' is chosen as one end of the vector and stand 'b' as the other. Now select the highest value from all the matrix value of the two stands. Subtract all of the matrix values of stand 'b' from the highest value selected above and add these values to the scores for the same stands in respect of stand 'a' and divide the sum by 2 to get an average value for each stand. Arrange the stands in order of the value so obtained, keeping stand 'a' at one end and stand 'b' at the other as the first axis.

In order to prepare the second axis subtract the lowest value on the first axis from the highest value of the same axis and divide the resultant value with 2. This value is the mid-point of the first axis. Examine all the stands and select those whose values on the first axis are ± 5 from the mid-point. Select thetwo most unlike stands (the two whose original matrix value is lowest) from the above. Repeat the same procedure as described for the first axis in selecting 'a' and 'b'.

Now on a graph sheet plot each stand with reference to the two axes. The horizontal axis is the first and the vertical axis is the second axis. The length of the axis should be equal on the graph sheet.

INTERPRETATION

Examine the curves obtained and see if any reasonable pattern (with high values together and a gradient from higher to lower) emerges. Can this pattern be correlated to the environmental factor?

Procedure C (Continuum): In a floristic province, the plant stands can be placed on a scale correlated with important environmental factors from one extreme to the other. In this concept it is assumed that plant communities are not discrete but they merge into each other showing a definite gradient which is correlated with the environmental gradient.

Determine relative frequency, relative density and relative dominance of each species in each stand and compute the importance value index, (Excercise 1.14/A). Select a few species, with high importance value, which are leading dominants in maximum number of stands studied. Now arrange the data in the following table.

AVERAGE IMPORTANCE VALUE OF SELECTED SPECIES IN STAND WITH GIVEN SPECIES AS LEADING DOMINANTS

No. of stands 1	Selected spp. 2	Leading dominant in stand 3 4 5 6 7
		A B C D Climax adaptation number

The average importance value is calculated by totalling up the importance value of the species in all the stands where that species is leading dominant and dividing the sum by the number of stands. Place the importance value of species A under the column 3 for all the stands dominated by A. In the column 4 place the importance value of A for the stands dominated by B and in column 5 for stands dominated by C and so on. Likewise arrange importance value of species B, C, and D.

The order in which these species are arranged in the above table is decided by some noted relationship with the environment. Generally climax species are put on the top and pioneer at the bottom. However, if this method is found untenable, the order is decided by trial and error placement until both horizontal and vertical columns most nearly approach a smooth curve on a graph.

In the next step a climax adaptation number is allotted to each species, giving number 10 for the climax species and 1 for the pioneer. Once the climax species is ascertained and number 10 is allotted to it, all other species can be rated on the basis of the distance of their optimum stands from those dominated by the climax species. Subdominant species are assigned number on the basis of their association with the leading dominants. More than one species may have the same climax adaptation number.

Next, the importance value of each sppecies is multiplied by its climax adaptation number in a particular stand. The resultant products (for all the species in the same stand) are added and the sum is called continuum index of that stand. This varies from 300 to 3000.

Calculate the average importance value of each species for each 100 unit segment of the continuum index and calculate the moving average (Brown and Curtis 1952). An example will make it clear. The average importance value of *Acer saccharum* in the 100 segment of continuum index from 2490 to 2590 was 92, for segment 2510 to 2610, 95 and from 2500 to 2600, 93. Using the following formula smoothed value was calculated.

$$B = \frac{n_1a + 2n_2b + n_3c}{n_1 + 2n_2 + n_3}$$

where B is the smoothed value of average b and n_1, n_2 and n_3 are the number of stands included in averages a, b, and c, respectively.

$$B = \frac{100 \times 92 + 200 \times 93 + 100 \times 95}{100 + 200 + 100} = 68 \pm$$

Plot these smoothed points on a graph sheet against continuum index for each of the leading dominants. In the series of smooth curves thus obtained, high points of each indicates its maximum importance and present a composite picture of the relationship of the species to each other.

INTERPRETATION

Plot a graph showing continuum index on horizontal and any of the environmental factors on vertical axis and see if any relation between the two gradients is exhibited.

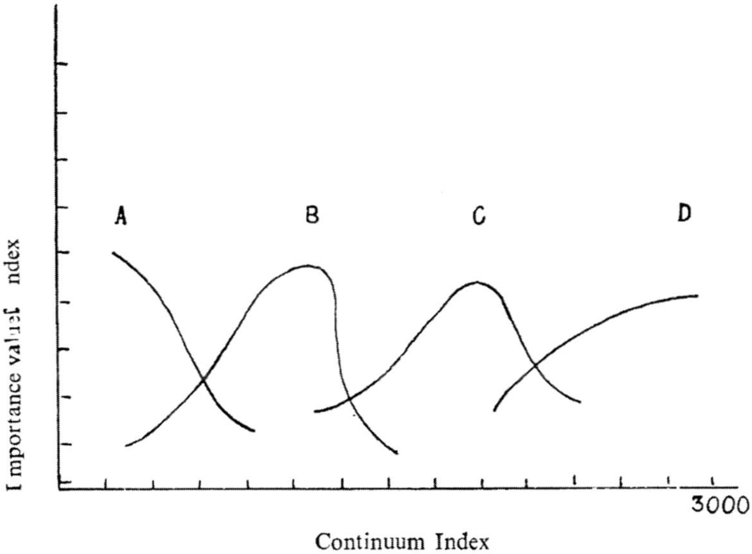

NOTE: In the place of Importance Value Index percentage cover as determined by Point—Frame method may also be used.

REFERENCES

Brown, R. T. and J. T. Curtis, 1952. The upland conifer-hardwood forests of nothern Wisconsin. *Ecol. Monogr.* **22**: 217-34.

Curtis, J. T. and R. P. McIntosh. 1951. The upland forest continuum in the prarie forest border region of Wisconsin. *Ecology* **32**: 476-96.

1.16 Charting and Mapping of Vegetation

Procedure A (Charting): Charting of a community gives an idea of spatial distribution of component species. This is done by determining the area occupied by each species and marking the same on a graph sheet (exercise 1.13/A). Thus prepared chart of the community represents the detailed structure, and is very useful to an understanding of the community dynamics. For this purpose the position of the quadrat should be precisely marked so that the subsequent charting is done on the same spot.

REFERENCES

Tansley, A. G. and T. F. Chipp. 1926. *Aims and Methods in the Study of Vegetation.* London.

Weaver, J. E. and F. E. Clements. 1938. *Plant Ecology.* McGraw Hill Book Co , New York.

Procedure B (Mapping): Make a general survey of the grassland to be mapped. Delimit the communities on the basis of dominants, and also note the soil types and topographic features. The extent of communities and the transition of one to another may also be determined. By giving suitable symbol to different plant species outline the areas covered by each community on a graph sheet using a suitable scale. The changes in the topography, substratum etc. may also be shown on the same map.

REFERENCE

Weaver, J. E. and F. E. Clements. 1938. *Plant Ecology.* McGraw Hill Book Co., New York.

1.17 Inter and Intra-specific Relations

I. INTERSPECIFIC ASSOCIATION

Procedure A: Association in the statistical sense, is the amount of co-occurrence in excess of that to be expected if two species are independently distributed. Positive association is to be anticipated in the case of the two species which exhibit overlapping habitat requirements or which interact in such a way as to favour mutual presence. Negative association in this sense is to be anticipated when habitat requirements or tolerances are distinct, or where through disoperative phenomenon, one species tends to exclude the other (Cole 1949). When a worker has collected data of presence and absence of any two species in a large number of sampling units at one or more sites, he can perform the test of association through Chi-square. The two prerequisites are: (1) a large number of sampling units and (2) the size of the sampling unit, enough to accommodate the test species together. The size of the sampling unit should always be mentioned along with the results.

The arrangement of a 2×2 contingency table for testing interspecific association is as follows:

		Species A		
		Present	Absent	
Species B	No. of times present	a	b	a+b
	No. of times absent	c	d	c+d
		a+c	b+d	a+b+c+d=n

The number of sampling units in which both the species A and B occur together is mentioned in the cell 'a' and the number of sampling units in which both the species are absent is inserted in the cell 'd'. In the cells 'b' and 'c', the number of sampling units having only species B or species A, respectively, are mentioned. Now the Chi-square value is calculated by the usual formula:

$$\chi^2 = \frac{(ad-bc)^2 n}{(a\times b)\ (a+c)\ (c+d)\ (b+d)}$$

If the χ^2 value thus obtained is significant at 5% level, according to the null hypothesis, i.e., "the two species are distributed independently" is discarded and the conclusion is drawn to the effect that both the species are associated, either positively or negatively.

Procedure B: Determine frequency of the plant species to be tested for association by laying more than 150 quadrats of requisite size. Now prepare the 2×2 contingency table as explained above and insert the different values. Calculate χ^2 value from the formula given above. Statistically significant value of χ^2 establishes that the two species under study are associated. Find out whether this association is positive or negative. For this expected values for each cell of the contingency table have to be calculated, assuming that both the species are independently distributed. Denote these expected values by a_1, b_1, c_1, d_1, respectively. Calculate through the following formulae:

$$a_1 = \frac{(a+b)\times(a+b)}{n}$$

$$b_1 = \frac{(b+d)\times(a+b)}{n}$$

$$c_1 = \frac{(a+c)\times(c+d)}{n}$$

$$d_1 = \frac{(c+d)\times(b+d)}{n}$$

Now compare the observed and expected values.

If the two species co-occur in more quadrats than expected (that is greater than a_1), and the individual occurrence of both the species is less than expected, the association is positive. If, however, the case is opposite the association is negative.

INTERPRETATION

What may be the cause of positive or negative association between the pair of species studied by you?

Procedure C (Degree of association): In the above procedure only the presence or absence of the interspecific association can be determined. To know the amount of association (or degree of association) Cole's index which uses the figures of 2×2 contingency table set above, can be used.

Construct 2×2 contingency table and calculate Cole's index from the following formulae.

(a) when ad $\geqslant$ bc

$$C \pm \delta c = \frac{ad-bc}{(a+b)\ (b+d)} \pm \frac{(a+c)\ (c+d)}{n(a+b)\ (b+d)}$$

c = standard error.

(b) when bc>ad and d $\geqslant$ a

$$C \pm \delta c = \frac{ad-bc}{(a+b)\ (a+c)} \pm \frac{(b+d)\ (c+d)}{n(a+b)\ (a+c)}$$

(c) when bc>ad and a>d

$$C \pm \delta c = \frac{ad-bc}{(b+d)\ (c+d)} \pm \frac{(a+b)\ (a+c)}{n(b+d)\ (c+d)}$$

INTERPRETATION

Explain the differences in the degree of association between any two pairs of species.

REFERENCES

Cole, L. C. 1949. The measurement of interspecific association. *Ecology* **30**: 411—24.

Goodall, D. W. 1953. Objective methods for the classification of vegetation: The use of positive interspecific correlation. *Austr. Jour. Bot.* **1**· 39—63.

II. COMPETITION

Procedure A (Competition): Plants, for their growth, depend upon the supplies available in the substrata they occupy. Thus from the same habitat a large number of individuals get the supply of water, nutrients and light. When the demand for these requirements in an ecosystem exceeds the availability, competition ensues resulting into depressed or modified growth of the competing individuals. The competition may be inter-or intra-specific depending upon whether the individuals of the same or different species are involved.

Fill earthenware pots of uniform size with equal amount of garden soil mixed with farmyard manure. Sow seeds of the plant species under study in the pots. After the emergence of seedlings, maintain five levels of plant density, viz., 1, 2, 4, 8 and 16 individuals per pot by thinning. Keep at least three replicates for each treatment. After the seed setting is complete, wash the pots and note the performance of plants under each treatment separately by recording the following growth parameters as indicated in the following table.

EFFECT OF CROWDING ON THE PERFORMANCE OF THE INDIVIDUALS

Growth parameters	No. of plants per pot				
	1	2	4	8	16
Plant height (cm)					
Plant spread (cm)					
*Leaf length (cm)					
*Leaf breadth (cm)					
Dry wt. (g) shoot/plant					
Dry wt. (g) root/plant					
Seed output per plant					

*Measurments to be taken of a mature leaf from a definite node in all the cases.

Plot the values for different growth parameters against the corresponding density on a graph paper.

INFERENCE

Derive the inferences regarding the effect of increasing density on the growth performance on the basis of graph curve and the data set in the table. Indicate what growth parameter is most affected by increasing density.

REFERENCE

Tripathi, R. S. 1965. *An Ecological Study of Weeds Infesting Wheat and Gram Crops of Varanasi.* Ph.D. Thesis, Banaras Hindu University.

Procedure B: Follow the method employed in procedure A and record the data periodically, say at 10 days' interval, instead of taking observations only at the end of the experiment. Record the observations in a table.

An examination of the data so collected will give an idea of the stage at which the effects of intraspecific competition on the individuals become observable.

Procedure C (Comparison of susceptibility): Follow the method employed in procedure A with a number of plant species of different life forms, say grasses and forbs, and record the data as shown in the following table.

Plant species	Growth parameter	No. of plants per pot					
		1	2	4	8	16	
Species A	Plant height (cm)						
	Plant spread (cm)						
	Dry weight (g) shoot/plant						
	Dry wt. (g) root/plant						
	Seed output/plant.						
Species B	Plant height (cm)						
	Plant spread (cm)						
	Dry wt. (g) shoot/plant						
	Dry wt. (g) root/plant						
	Seed output/plant						
Species C	Plant height (cm)						
	Plant spread (cm)						
	Dry wt. (g) shoot/plant						
	Dry wt. (g) root/plant						
	Seed output/plant						

Plot the values for individual growth characters of all the species on X axis against the density on Y axis.

INFERENCE

Compare the results obtained in case of different grassland species and note the order of susceptibility to intraspecific competition. Also note the pattern of distribution of the selected individuals in grasslands and mark if it has any relation with the susceptibility of the species to intraspecific competition.

REFERENCE

Clements, F. E., J. E. Weaver and H. C. Hanson. 1929. *Plant Competition—An Analysis of Community Function*. Carnegie Institute, Washington, Publ. No. 398.

Procedure D (Comparison of competitive ability): The extent of adverse effect of competition depends upon the competitive ability of a plant 'which' according to Welbank (1963) 'can be thought of as its ability to depress the growth of other plants. This will be determined by its size and vigour, and so will depend on what species it is competing against (because this competitor will have depressive effect), as well as on its own intrinsic characteristics'. This can be measured by employing a simple technique in which the effect of test species on a common plant species (indicator) is noted. This simple method is called indicator method. Usually the number of plants of the test species in a pot is kept high against a single plant of indicator species (Welbank 1963).

Take earthenware pots of uniform size filled with equal amount of garden soil. Allow the seedlings of the indicator, and of the species of which competitive ability is to be measured, to come up in the pots. By thinning maintain the density per pot as shown below:—

1. Ten individuals of test species+one indicator plant.
2. Single indicator plant (control).

All the above treatments should be replicated at least thrice. After a period of 2 or 3 months wash the pots and take dry weight (an index of competitive ability) of shoots and roots of indicator plant in all the cases including control. Record your data as shown in the following table.

EFFECT OF TEST SPECIES ON DRY WEIGHT OF SHOOT AND ROOT OF INDICATOR PLANT

Treatment		Dry weight of shoot per plant (g)	Dry wt. of root per plant (g)
Indicator plant growing alone			
Indicator plant growing with species	A		
,, ,, ,, ,,	B		
,, ,, ,, ,,	C		
,, ,, ,, ,,	D		
,, ,, ,, ,,	E		
,, ,, ,, ,,	F		

INFERENCE

Note which test species reduced both/either the tops and/or roots of indicator plant most, and which the least. The species causing least reduction possesses the poorest competitive ability and the species which reduces the dry matter production of the indicator plant most, has the highest competitive ability.

Repeat the above experiment using some other indicator plant and see whether the order of competitive ability of the test species changes.

N.B.— For indicator plant, any annual plant most adapted to local conditions may be selected.

REFERENCES

Tripathi, R. S. 1965. *An Ecological Study of Weeds Infesting Wheat and Gram Crops of Varanasi.* Ph.D. Thesis, Banaras Hindu University.

Welbank, P. J. 1963. A comparison of competitive effects of some common weed species. *Ann. Appl. Biol.* **51:** 107—25.

Procedure E (Interspecific competition): Raise seedlings of two different plant species in a glasshouse and transplant in earthenware pots. Maintain two sets of experiment. In one, keep the density per pot of the species 'A' constant, and vary the density of species 'B'; and in another set, do the reverse, as indicated in the tables. After a period of 8 or 10 weeks, wash the pots under each treatment, and take oven dry weight of the plants of both the species separately. Use the following tables for recording the data.

DRY WEIGHT OF THE PLANTS OF SPECIES 'A' AND SPECIES 'B' UNDER INTERSPECIFIC COMPETITION
(No. of individuals of species 'A' per pot constant and of species 'B' variable).

No. of individuals per pot		Dry weight of plants per pot	
Species 'A'	Species 'B'	Species 'A'	Species 'B'
2	0		
0	1		
2	1		
0	2		
2	2		
0	4		
2	4		
0	6		
2	6		
0	8		
2	8		

DRY WEIGHT OF THE PLANTS OF SPECIES 'A' AND SPECIES 'B' UNDER INTERSPECIFIC COMPETITION
(No. of individuals of species 'B' per pot constant and of species 'A' variable)

No. of individuals per pot		Dry wt. of the plants per pot	
Species 'A'	Species 'B'	Species 'A'	Species 'B'
0	2		
1	0		
1	2		
2	0		
2	2		
4	0		
4	2		
6	0		
6	2		
8	0		
8	2		

CALCULATIONS

Calculate in each case the percentage reduction in dry weight of competing species by using the following expression:

$$\% \text{ reduction} = \frac{x-y}{x} \times 100$$

where y = dry weight per pot of a particular species when growing in competition with another species and x = the same without competition.

Put in a separate table the % reduction in dry matter production of species 'A' and species 'B'

PERCENTAGE REDUCTION IN DRY MATTER PRODUCTION OF SPECIES "A" DUE TO PRESENCE OF SPECIES "B" AND VICE VERSA

No. of plants per pot		% reduction in dry matter production	
Species 'A'	Species 'B'	Species 'A'	Species 'B'
2	1		
2	2		
2	4		
2	6		
2	8		

INTERPRETATION

What is the effect of increasing density of species 'B' on a constant density of species 'A' and *vice versa*?

Note whether the reduction in yield of species 'A' by species 'B' is more, or of species 'B' by species 'A', and comment upon the relative success of the individuals of the two species. In case the reducing effect of one species on another is very marked, what is the situation in a grassland—whether the two grow side by side or are distributed apart from each other?

REFERENCES

Mann, H. H. and T. W. Barnes. 1947. The competition between barley and certain weeds under controlled conditions. II. Competition with *Holcus mollis*. *Ann. Appl. Biol.* **34:** 252—66.

Mann, H. H. and T. W. Barnes. 1950. The competition between barley and certain weeds under controlled conditions. IV. Competition with *Stellaria media*. *Ann. Appl. Biol.* **37:** 139—48.

Procedure F (Comparison of inter-and intra-specific competition): Select a number of pairs of treatments (shown in tables of procedure E) with equal number of individuals per pot, of each selected pair. One set should contain the individuals of the same plant species and the other set should contain the individuals of both the species. Calculate the dry matter production per plant and compare the effects of intra- and interspecific competition.

INFERENCE

Note the type of competition which has got more adverse effects on the species concerned.

Procedure G (Shoot and root competition): Divide the pot internally into two equal halves with the help of a metal or glass plate, which extends from the soil surface down to the bottom of the pots. Fill each half with soil to the same level. Raise seedlings of species 'A' in one half, and of species 'B' in the other half. After the seedlings are well established, maintain suitable and equal density of each species in each half by thinning. In this case shoots of the two species are allowed to come into contact with each other but the roots are kept separated. Thus only shoot competition takes place.

FIG. 1.17/II/G

For the study of root competition, shoots of the two species are separated by means of metal plates of appropriate size which extend from the soil surface upwards (as shown in figure) while the roots are allowed to intermingle freely.

Determine the dry weight of shoots and roots of the plants after a period of 2-3 months. Record the data in the table.

THE EFFECT OF SHOOT AND ROOT COMPETITION ON THE DRY WEIGHT OF SHOOT AND ROOT OF SPECIES 'A' AND SPECIES 'B'

Plant species	Shoot competition		Root competition	
	Dry wt. shoot/plant (g)	Dry wt. root/plant (g)	Dry wt. shoot/plant (g)	Dry wt.root/plant (g)
Species 'A'				
Species 'B'				

Present the data in a histogram as shown in the following figure.

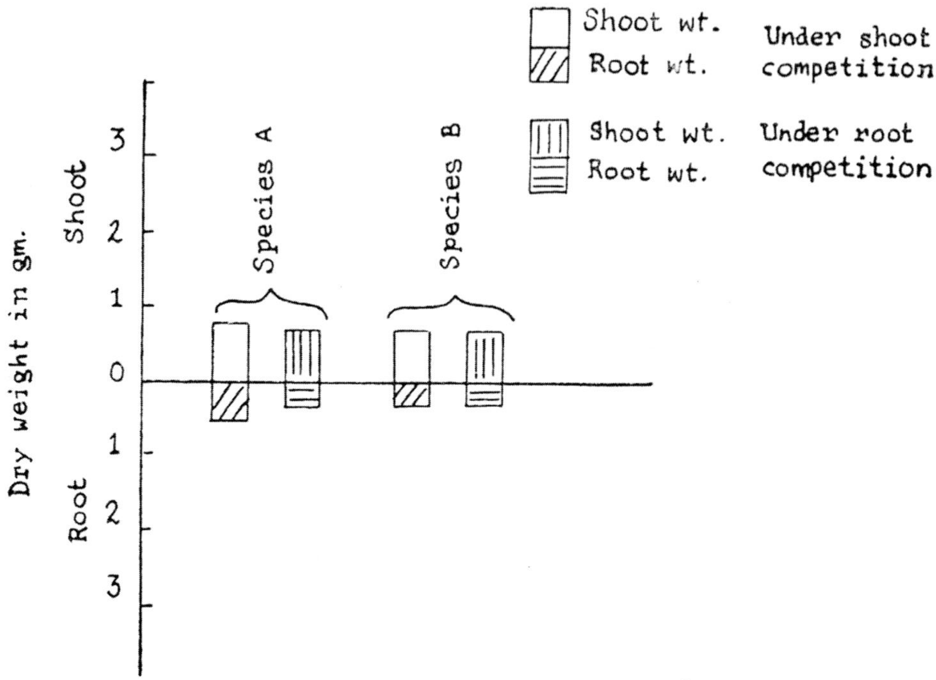

Fig. 1.17/II/G Histogram showing the effect of shoot and root competition between species A & B.

INTERPRETATION

Mark if there exists any significant difference in dry weight of shoot and/or root of a species growing under shoot and root competition.

REFERENCE

Wilson, D. B. 1962. Seedling growth of two grasses at different soil temperature and under different types of competition. *Advancing Frontiers of Plant Sciences*. **1**: 211—218.

1.19 Profile Diagram

Sometimes there is clear cut stratification, i.e. different plant species occupy different strata. In a grassland, at least two strata, one of spreading species and the other of vertically growing species are quite distinct. The bisect and the profile studies are quite useful for studying layering or stratification in a community. Besides being applicable to aboveground vegetation, these may also be equally applied to root studies. The idea of relative position and the extent of penetration of the root system of different species is often very helpful to an understanding of moisture and nutrient relations of the species growing together. This information may help to some extent to explain whether the competition among the associates will be severe or otherwise.

Select a piece of grassland and dig a trench deep enough to ensure the inclusion of deep penetrating root systems. Carefully excavate the root systems and determine the position and extent of main and lateral roots. Giving a suitable scale on a graph-sheet, draw the root systems of different plant species. This will give the idea of stratification of underground parts.

On the same graph sheet profile (stratification and lateral spread) of the above-ground parts may also be shown.

INTERPRETATION

(1) Name the shallow rooted plant species, and the species whose root system penetrates deeper than 30 cm.

(2) Find out the percentage of shallow rooted annuals and compare with the percentage of such perennials. Also note whether the majority of the deep rooted plant species are perennials.

(3) List the plant species forming the lowest layer, i.e. the spreading species.

(4) Name the dominant species which grow higher and form the next storey of vegetation.

REFERENCE

Oosting, H. J. 1958. *The Study of Plant Communities.* W. H. Freeman & Co., San Francisco.

Weaver, J. E. and F. E. Clements. 1938. *Plant Ecology.* McGraw Hill Book Co., New York.

1.20 BEHAVIOUR OF COMPONENT SPECIES (AUTECOLOGY) OF THE ECOSYSTEM

The community is constituted of populations (species) of different kinds, and each population influences the ecosystem according to its ecological life-history.

1.21 Geography, Evolution and Systematics

I. GEOGRAPHY AND EVOLUTION

Plant species are widely distributed over the surface of the earth. Information regarding the occurrence of any species in different regions of the world has to be obtained from botanical gardens, herbaria and publications. This information should be compiled in a table. An example is given in the following table for the distribution of a number of species of *Stipa*.

WORLD DISTRIBUTION OF A NUMBER OF SPECIES OF *Stipa*

Region	Number of species reported
Australia[1]	69
New Zealand[3]	69
Tasmania[3]	2
Asia[2] (Japan 2; China 13; USSR 60; India, Burma and Pakistan 27; and Afghanistan 7)	100
South Africa[3]	3
United States[4]	14

1. Information obtained from the Botanist, Royal Botanic Gardens, Sydney.
2. Information obtained from the Director, Royal Botanic Gardens, Kew, and subsequent examination of the specimens at the Herbarium.
3. Information obtained through Index Kewensis (Hooker and Jackson 1946).
4. Information compiled from Hitch-Cock (1925).

Similarly the distribution of the species can be shown in a map of the world.

Consideration of the present and past geographical distribution of various members of the taxon has helped in plotting the dispersal trends and migration routes of the plant species. The distribution pattern throughout the world provides a foundation for the proposed origin of areas and the centres of secondary dispersal of the group.

There are genetically distinct micropopulations within a species, which on account of interfertility remain under the same taxonomic unit. But habitat selections may lead to ecotypic differentiation.

For example, *Lindenbergia polyantha* and *L. urticaefolia* are shown by Misra and Rao (1948) to be ecotypes of the same species. They interbreed freely and produce intergrading forms but remain isolated in nature as *L. polyantha* can tolerate a high dose of lime on walls. Ramakrishnan (1961) has shown that *Euphorbia thymifolia* has a copper coloured ecotype which is a facultative calciole population, and a green coloured ecotype, which is an obligate calcifuge population of the same species. In this case, as will be seen, taxonomists have not erected two different species as was done in *Lindenbergia urticaefolia*.

REFERENCES

Misra, K. C. 1963. Phytogeography of the genus *Stipa*. *Trop. Ecol.* **4** : 1—20.

Misra, R. and B. S. Shiva Rao. 1948. A study in the autecology of *Lindenbergia polyantha* Royle. *J. Indian bot. Soc.* **27** : 186—199.

Ramakrishnan, P. S. 1961. Calcicole and calcifuge problem in *Euphorbia thymifolia* Linn. *J. Indian bot. Soc.* **40** : 66—81.

II. SYSTEMATICS

Systematics deal with the description, identification and naming of plants, and their classification into different groups according to their resemblances and differences, mainly in their morphological characteristics. The plants are numerous, and are of varied types, and hence require a systematic study.

Taxonomists' description of the species, given in published flora, is based on herbarium studies depending upon the limitations of time and sampling of the field specimens. An ecologist while using the text, carefully records variations in structural and behavioural characters of the populations as found in different ecosystems. Field studies thus lead to formulation of ecological hypotheses which can be tested experimentally.

III. DISPERSAL

It is through dispersal of their propagules that the plant species get spread from one place to another. Dispersal is brought about in a variety of ways. Light seeds and the seeds provided with parachute mechanism, can easily be blown from place to place, while seeds which can float on water may be dispersed through water currents. Apart from these natural agencies, dispersal also takes place through animals and human agency. In a grassland community, the grazing animals, like goat, sheep and cow play an important role in dispersal. Removal of the herbage from grassland for using as fodder helps in the dispersal of the seeds to distant places. Adaptations which ensure the dispersal through biotic means are: the capability of the seeds to retain their viability even after passing through alimentary canal of animals, the palatability of fruits which attract birds, rodents and ants which carry them to distant places; the presence of hooks, spines, awn or retrorse hairs on fruits and seeds to facilitate their attachment to hairy skin of animals, feather of birds and to clothes of man.

Introduce a cow, sheep or goat in a grassland. After a few hours examine thoroughly the body of grazing animals, collect the disseminules attached to the hairy skin, and count them. Also collect in a tray the animal droppings from a grassland and subject it to favourable conditions for germination of the contained seeds. Identify and count the seedlings coming up in the tray from time to time.

For examining the extent of dispersal through human agency, traverse through the grassland and count the propagules collected in your shoes, and also those attached to your trousers.

INTERPRETATION

(1) Note whether the seedlings coming up in the tray are the typical components of the grassland under study.

(2) Give an account of the nature of the seeds attached to your clothes, and point out in which respects these differ from those collected in shoes.

REFERENCES

Oosting, H. J. 1958. *The Study of Plant Communities.* W. H. Freeman & Co., San Francisco.

Ridley, H. N. 1930. *The Dispersal of Plants Throughout the World.* L. Reeve & Co. Ltd., London.

1.22 Structural and Physiological Characters and Their Variations

I. STRUCTURAL CHARACTERS

Most classifications have employed morphological and anatomical characters which are quite variable in relation to the habitat factors. Qualitative and quantitative variability may be recorded from individuals which have grown in different habitats.

Procedure A (General): In order to study morphological and anatomical variations, collect plants from at least 3 uniform populations from each of the localities representing the range of habitats. From each population 30 plants be collected. For the study of the flower at least 5 flowers from each plant be taken. The data be recorded in a table.

MEASUREMENT OF MORPHOLOGICAL AND ANATOMICAL CHARACTERS OF THE SPECIES FROM DIFFERENT HABITATS IN NAMED LOCALITIES

Characters	Habitats					
	A	B	C	D	E	F
A. Morphological characters						
(1) Root length						
(2) Root diameter						
(3) Shoot length						
(4) Shoot diameter						
(5) Length of the 5th and the 8th internode from apex or base						
(6) Length of leaf (at 5th internode)						
(7) Breadth of leaf (-do-)						
(8) Floral characters						
(*a*) Size of sepal						
(*b*) Size of petal						
(*c*) Size of the stamen						
(*d*) Size of the carpel						
(9) Colouration, its depth and pattern including those of green parts						
(*Note*—See exercise 1.25 for characters of fruit and seed)						
B. Anatomical characters						
(1) Area ratio of vascular/ground tissue						
(2) Area of cross section of aerenchyma, if any						
(3) Size of air spaces, if present						
(4) Any other anatomical character which may be related to the habitat—such as stomatal frequency* and index*, thickness of cuticle, thickness of palisade layer, hairs, etc.						

NOTE: In all anatomical studies material of the same age should be selected. If the data are statistically analysed, give significant results that may predict the natural habitat or adaptation, and genetical variations.

*The leaf may be stained with gentian violet dissolved in 90% alcohol. This will result in clear demarcation of the guard cells from the other epidermal cells. The epidermis is then peeled carefully by a pair of forceps. If removal of epidermal peelings is not possible, the leaf surface is coated with collodion. The thin film, when dry, is removed and stomata are counted from the impression on the under surface of the leaf.

From the above table, those characters which vary from habitat to habitat, have to be taken in separate tables and analysed statistically. Questions and problems are then formulated for experimental testing.

Procedure B (Size of stomatal aperture): (1) Instead of counting the number of stomates, per unit area of the peeling, measure the size of the stomates by the micrometer.

(2) A series of infiltration solutions of different concentrations (Isopropyl alcohol or Isobutyl alcohol or Xylol) is applied. The discolouration of the other side proves the infiltration of the solution in question. The infiltration of lower concentration expresses for large stomata and that of higher concentration for the small ones.

Procedure C (Root investigation): Collect the roots of the plant from different localities and habitats by the following methods—

(1) Bisect method
(2) Monolith method
(3) Simple digging

Note the morphology of the roots and record their length and diameter.

Name of species..........................

Locality	Habitat notes	Morphological notes	Length	Diameter

Chart the root system on a graph paper while studying by trench method and show the distribution of the branches in different layers of the soil.

Take the fresh weight of roots and put them for drying at 65°C. The dry weight of the roots is then recorded. Determine the percentage moisture of the root tissues and the soil separately.

Station........................
Date...........................

Locality and habitat notes	Fresh weight	Dry weight	Percentage moisture

II. PHYSIOLOGICAL CHARACTERS

Procedure A (Habit and phenology): Habit and phenological observations on the plants are noted seasonally, and recorded as below :

Station................ Date and time, where necessary

1. Habitat
2. Seedling emergence
3. Leaves attaining full size
4. Elongation of shoot ceases
5. Flower bud initiation
6. Flowering starts
7. Fruiting starts
8. Seed dispersal
9. Leaves begin to die
10. Leaves begin to fall
11. All the leaves have fallen
12. Complete death in annuals

Habit and phenology of a plant are as much expressions of the genotype, in response to the environment, as other structural characters. If they differ in different localities, biotypes need further investigation.

Procedure B (Root/shoot ratio): Collect the roots and shoots of individual plants of the species from different localities, representing different habitats, and determine the ratio on dry weight basis. Tabulate the data as given below.

Date..........................

Locality and habitat notes	Root			Shoot			Root/ shoot ratio
	Fresh wt.	Dry wt.	moisture %	Fresh wt.	Dry wt.	moisture %	

The above study helps in understanding the relative growth of root and shoot in the sun, shade and moist or dry soil, etc. If the plants examined are of the same age, the effect of these factors on production of dry matter can also be inferred.

Procedure C (Osmotic pressure of sap): Samples collected should be frozen and stored at -20°C until required. The sap is extracted by a press applying constant pressure for all samples. The sap is poured in the freezing chamber embedded in the freezing mixture and is stirred continuously. The freezing temperature is noted by Beckman thermometer.

Calculations: $OP = 12.06\Delta - 0.021\ \Delta^2$ where Δ is the freezing point depression.

Osmotic pressure is expressed in terms of atmospheres, and can be computed to any temperature using gas laws.

Procedure D (Photoperiodism): Day-length of a particular value (*critical day-length*) which differs widely from species to species, is necessary for initiation of flowering. Plants in which flower initiation is promoted by day-length shorter than the critical day-length are called "Short-day plants" while those requiring day-length longer than a particular value are called "Long-day plants". Plants which flower apparently irrespective of day-length are called "Day-neutral" or "Day-length-indifferent plants".

Transplant young seedlings (5 or so per pot) at cotyledonary stage in the flower pots and expose them to day-length varying from 5 to 24 hrs per day. The natural day-length may be supplemented by a 500W incandescent lamp. The plants should be removed to darkness after the required light period.

The experiment may be performed by using at least three replicates. All other environmental conditions should be maintained identically.

Record the following data and time required for the first appearance of the floral stage in various treatments.

RECORD OF GROWTH UNDER DIFFERENT DAY-LENGTHS

Date of observation (weekly or fortnightly)	Day-length	Height of the plant	No. of branches	No. of leaves	Date of flowering
	5 hrs.				
	7 ,,				
	9 ,,				
	11 ,,				
	13 ,,				
	16 ,,				
	20 ,,				
	24 ,,				

Find out the dry weight of the root and shoot, and calculate shoot/root ratio.

INFERENCES

(1) What is the critical day-length of the species?

(2) What is the performance of plants under shorter or longer day-lengths?

(3) Can this information explain the geographical distribution of the species?

(4) What happens (in case of short-day plants) if the plants are illuminated in the middle of the dark period?

(5) Collect the seeds formed under different photoperiods and find out the differences in the germination behaviour.

(6) What are the main theories propounded for the control of photoperiodism?

Read about Phytochrome theory, Florigen theory, Balance between vegetative growth and reproductive growth.

REFERENCES

Hillman, W. S. 1964. *The Physiology of Flowering*. Holt, Rinehart and Winston, New York.

Withrow, R. B. (Ed.) 1959. *Photoperiodism and Related Phenomena in Plants and Animals*. American Assoc. for the Advancement of Science, Washington, D. C..

Procedure E (Thermoperiodism) :

(i) Find out the upper and lower temperature limits during the growth period of the plant.

(ii) Grow the plant at constant high and low temperatures separately.

(iii) Grow the plant at alternate temperatures of varying periods.

(*a*) High temperature during light period.

(*b*) Low temperature during light period.

Record the growth performance as in case of photoperiodism.

NOTE: It is generally seen that a thermoperiodic plant requires low nocturnal temperature.

REFERENCES

Went, F. W. 1957. *The Experimental Control of Plant Growth*. Chronica Botanica, Waltham, Mass.

————.1961. Temperature. *Encyclopaedia of Plant Physiology* **16:** 1—23.

1.23 Population Characteristics and Ecotypic Differentiation

A species is treated as a population of similar individuals, which will not interbreed freely with those of the other species. Most of the widely distributed species, however, consist of micropopulations based on small genetic differences and, therefore, capable of interbreeding. Such of these biotypes as are structurally, physiologically and developmentally adapted to the habitat conditions, are known as ecotypes. The micropopulations remain differentiated and isolated in nature on account of habitat differences, such as photoperiod, aridity, soil calcium, nitrogen, grazing, etc. Since many of the morphological characters are linked with those of physiological and developmental origin, a close examination of the populations in different habitats generally reveals the ecotypes. These can be tested either by statistical analysis, or by transplant experiments, for the phenotypic variations which are continuous against the genotypic ones which are discontinuous from population to population.

Procedure A: Suppose three main forms of leaves are found in a plant species. Each type may be graded into say, 6 different subtypes. Each individual plant may show, within a range, one of these types of leaves, depending upon its genotypic composition and the environmental conditions.

Collect leaves from plants of a species growing in different habitats, and make frequency classes as convenient from the leaf character.

VARIATION IN LEAF CHARACTER FROM DIFFERENT HABITATS

Habitats	No. of plants studied	Leaf length class	Leaf breadth class
		1 2 3 4 5	1 2 3 4 5
1	30		
2	30		
3	30		
4	30		
5	30		
Mean square for within population variation			
Variance Ratio			
Level of significance			

From the above table prepare graphs as in Fig. 1.23/A.

This type of unimodal curves give an idea of homogeneous population and there is no heterogeneity in its structure. Hence, possibly all the individuals belong to the same biotype.

If the figure comes out like that of graph II i.e. bimodal, there is a possibility of two different populations. A trimodal curve, as in graph III, is suggestive of the existence of three populations.

Precaution should be taken to select plants and leaves of the same age for measurements.

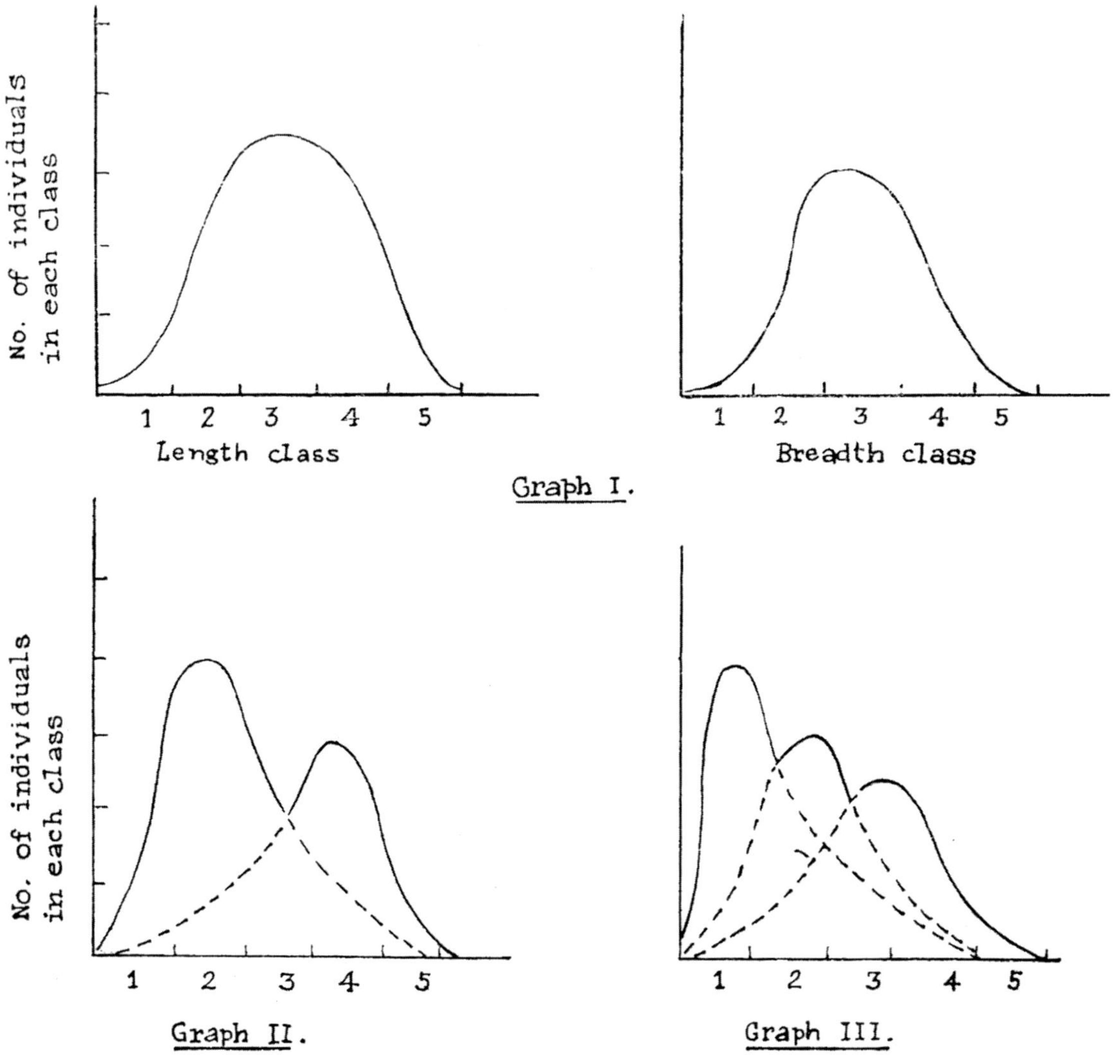

FIG. 1.23/A

Procedure B (Scatter diagram): Anderson (1949) has devised a method by which the populations can easily be differentiated. Two measurable characters are taken for this purpose, e.g. leaf length and leaf breadth, etc., and a large number of measurements are made from different localities. These are plotted on a graph by taking the two characters on X and Y axes. In case the population is heterogeneous, the various distinct populations will be separated out by showing aggregation of points at distinct places on the graph. In a homogeneous population, however, only one aggregation will be formed.

REFERENCE

Anderson, E. 1949. *Introgressive Hybridization.* John Wiley & Sons Inc., New York.

Procedure C: Turesson (1922) wanted to know if the differences in the apparent variants of the populations were due to modifications of the same genotype as a result of direct environmental influences or due to the expression of different genotypic combinations in different habitats. He performed a series of transplant experiments to clarify this aspect.

Some experiments performed at Varanasi, with species of *Euphorbia* are described. The following three populations were recognised in the field:

(*a*) Moist and shady habitat—erect and sparsely branched.

(*b*) Grasslands—prostrate and profusely branched.

(*c*) Foot paths—compact, small leaved and cushion formed.

Young seedlings of the same age from the 3 populations are transplanted in the garden (i) under identical conditions (neutral habitat) and also (ii) in each others' habitats (reciprocal habitats).

The footpath form grows into the same form as that of the grassland in the neutral habitat, and vice versa, in the reciprocal transplants.

INFERENCE

The two forms are genetically identical, so the differences disappear when they are grown under similar conditions or one form is transformed into the other in the reciprocal habitats.

The moist form maintains its erect sparsely branched nature as distinct from the prostrate grassland form, in both the neutral and reciprocal transplants, although their performance is better in their own media.

The grassland and the moist habitat ecotypes are established.

REFERENCES

Ramakrishnan, P. S. 1959. *Contributions to the Ecological Flora of Varanasi Distt.* Ph.D. Thesis, Banaras Hindu University.

Turesson, G. 1922. The genotypical responses of the plant species to habitat. *Hereditas* **31:** 211—350.

1.24 Cytogenetical and Breeding Behaviour

The ecotypes may be examined for their karyotype or meiotic behaviour, which may reveal cytological differences or/and subjected to breeding experiments.

Procedure A (Hybridization): The genetic behaviour of the ecotypes of *Euphorbia thymifolia* is revealed by the following breeding experiment.

Make a cross (hybridization) between the red form and the green form.

F_1 progeny is intermediate between the two parents. Undersurface of the leaf is reddish in colour and the rest of the plant body is green. In F_2 three intermediate forms are recognisable. The 5 forms starting with red form (RR) and gradually passing on to the green form (rr), segregate in the ratio of

$$\begin{matrix}(RR) & (Rr) & (Rr) & (Rr) & (rr) \\ (RR) & (RR) & (rR) & (rr) & (rr)\end{matrix} = 1:4:6:4:1$$

The two ecotypes interbreed freely. The inheritance of the red colour is governed by a duplicate factor. In the nature the

$$\begin{matrix}(RR) \\ (RR)\end{matrix} \text{ and } \begin{matrix}(Rr) \\ (RR)\end{matrix}$$

forms are found to grow in calcareous as well as non-calcareous soils, and the other three are restricted to non-calcareous soils only.

REFERENCE

Ramakrishnan, P. S. 1961. Calcicole and calcifuge problem in *Euphorbia thymifolia* Linn. *J. Indian bot. Soc.* **40:** 66-81.

Procedure B (Pollination): Pollination is the transference of pollen grains from the anther of a flower to the stigma of the same flower or of another flower of the same or sometimes allied species, often through various agencies. Pollination is of two kinds viz., (i) Self pollination or autogamy (ii) Cross pollination or allogamy.

One cannot study the pollination mechanism of flowering plants without being astounded by the variety of elaborate devices, which have evolved. Not only are flowers adapted structurally to secure this end, but in bisexual flowers, the anthers are commonly shed before the stigma is exposed or receptive (protandry), or in some cases the stigma matures first (protogyny).

Pollination need not always result in fertilization and fruit formation. Self incompatability leads to hybrid vigour. Outbreeding and inbreeding populations have suitable barriers for limiting them. Sterility and seed setting without fertilization are of wide occurrence in seemingly well pollinated flowers.

Enclose flowers from the bud stage in fine muslin or cellophane bags. Keep them for sometime and observe.

OBSERVATIONS

It is observed that little or no seed has set, confirming that the flowers are largely self-sterile.

There are some species which are habitually self pollinating, but even among these, self pollination may be restricted to the later formed flowers. *Commelina bengalensis* and some species of *Viola* besides many others are self-fertilized, and set seed without the buds ever opening.

In contrast to self-sterility, shown by most species when the flowers are 'bagged' to prevent cross pollination, counts of the percentage seed-set under normal conditions usually point to a high degree of efficiency in the pollination mechanism.

1.25 Reproductive Capacity, Dormancy, etc.

I. SEED OUTPUT

A population perpetuates, multiplies and disperses through propagules, which are seeds in most of the species. Hence the number of seeds produced by a plant gives a rough estimation of its potentialities for increase. It also shows the amount of energy passed by a species to the next generation.

Go to the field at the time of seed setting and collect specimens by "random sampling". If the plants show variations, each type should be collected in separate lots.

Plants growing in different light, moisture and soil conditions at a certain locality should not be mixed up (why?). In some species, the seed formation is periodic. In such cases the plants should be tagged and seeds collected at the proper time before they are shed. Similarly if the dispersal is quick, the plants or capsules should be tied with a cloth or polythene bag to trap the seeds.

The seed production of plants, where the number is not too great, may be determined directly by actual counts of the seed produced.

The results should be statistically sound. Hence observations may be recorded for at least 30 plants or more.

In species where fruits contain more than one seed, the mean value for both the number of seeds per fruit and the number of fruits per plant should be determined.

OBSERVATIONS

Record your data as under:

RECORD OF NUMBER OF CAPSULES PER PLANT

Serial No. (plants)	Locality						
	1	2	3	4	5	etc.	total
1 2 3 and so on							
Total							
Arithmetic mean	$m \pm s$	$m \pm s$	$m \pm s$	$m \pm s$	$m \pm s$	$m \pm s$ $m \pm s$	$m \pm s$

Calculate standard error of mean.

RECORD OF NUMBER OF SEEDS PER CAPSULE

Serial No. (capsules)	Locality							
	1	2	3	4	5	6	etc.	Total
1 2 3 4 and so on								
Total								
Mean $\pm$ standard deviation	$m \pm s$							

Calcuate standard error of mean.

The average seed output

$=$The mean of the products of the two mean values
(No. of seeds per fruit and No. of fruits per plant)
$\pm$standard error of means.

NOTE: Salisbury (1942) has used 'Probable Error' for expressing average seed output. Moroney (1962) says that the term Probable Error was used in former times to denote 0.6745 s.d., this being the deviation just as likely to be exceeded as not. The term is now very much obsolescent. Hence the results may be expressed in Mean $\pm$ standard error of mean.

Analyse the data statisticaly.

Localities	Mean (seeds per capsule)	Standard deviation	Standard error of mean	Coefficient of variation
1 2 3 4 5 etc.				

Localities	Differences of means	Standard error of difference
1—2 or 1—3 2—3 3—4 and so on		

INTERPRETATION AND INFERENCE

(1) The individuals of the population vary in their seed output within a certain range.

(2) Compare the average seed output of plants from different localities. Is the variation significant? If so, is it related to some factor of the environment?

(3) Compare the seed output of the morphologically distinct individuals, in the same locality.

Are the results significant? If it is so, how will you ascertain whether the difference is due to environmental or genetical factors?

(4) What is the area covered by a mature plant? What will happen if all the seeds produced, develop into mature plants each year?

(5) Has the large seed output a survival value for the species in the ecosystem? Do they serve as food to other organisms?

REFERENCES

Moroney, M. J. 1956. *Facts from Figures.* Pelican Book.

Salisbury, E. J. 1942. *The Reproductive Capacity of Plants.* G. Bell and Sons Ltd., London. pp. 244.

II. SEED WEIGHT

The seed weight is a rough approximation to the supply of potential energy available for the seedling. The offsprings of different populations of a community face competition at the initial stage. The seed weight gives an idea of food reserve available to the embryo for its growth and development in absence of suitable environmental conditions at its early stage.

Have at least 30 lots of 10, 25, 50 or 100 seeds according to convenience and weigh them.

OBSERVATIONS

WEIGHT OF SEEDS (IN LOTS OF 100/50 ETC.)

No. of readings	Locality				
	1	2	3	4	and so on
1 2 3 4 and so on					
Total					

Mean$\pm$S.D.
Analyse the data statistically.

INTERPRETATION AND INFERENCE

(1) What may be the causes if the seed weight varies significantly?

(2) Are all the seeds of the same fruit equal in weight?

(3) Do the seeds formed at different nodes vary in weight?

(4) Do the seeds collected from the same plant at regular intervals vary in weight?

(5) Measure the size of the seeds (with the help of a binocular dissecting microscope and eyepiece micrometer if the seeds are small). Note if the smaller size and weight of the seed effect wide dispersal.

REFERENCE

Salisbury, E. J. 1942. *The Reproductive Capacity of Plants.* G. Bell & Sons Ltd., London.

III. VIABILITY

All the seeds produced by a piant may not have a well formed embryo to resume growth; or the embryo inside the seed has a definite life span under the given environmental conditions, which may determine the spread of the species within the ecosystem, with respect to time.

Procedure A: Cut the seed at one side and dissect out the embryo. Allow the embryo to remain between 2 pieces of filter paper in a petridish for a few days (at suitable temperature).

OBSERVATION

Record the length of the embryo.

INFERENCE

If the embryos develop, the seeds are viable, whereas non-viable embryos will deteriorate.

Procedure B: The viable seeds respire, which causes colourless tetrazolium dyes to change into highly coloured compounds by chemical reduction. The reaction does not occur if the seed is dead. The living tissue becomes red while non-living tissue remains uncoloured.

Prepare 0.1% solution of T. T. C. (2, 3, 5—triphenyl tetrazolium chloride) in distilled water. Cut each seed in half, longitudinally through the centre of the embryo. Immerse the halves in the above solution in a petridish and put in the dark for a few hours at pH between 6-7.

OBSERVATION

Record the number of seeds in which the embryos are stained carmine red. Calculate the percentage of stained embryos.

NOTE: Store the seeds in various temperature and moisture conditions, and test for the viability at different intervals. Does a positive tetrazolium test definitely point out that the seeds will germinate if optimum conditions are provided?

REFERENCES

Machlis, L. and J. G. Torrey. 1956. *Plants in Action*. W. H. Freeman & Co., San Francisco.

Hartmann, H. T. and D. E. Kester. 1959. *Plant Propagation*. Prentice-Hall Inc., Engelwood Cliffs, N.J.

IV. GERMINATION

Seed germination may be defined as the resumption of active growth by the embryo. Protrusion of the radicle through the seed coat is usually taken as a criterion for germination. The process is dependent upon a number of environmental factors, e.g., moisture, temperature, light, gaseous exchange, etc.

Since the species vary in their requirements for germination, the experiments help to explain their distribution in space and time.

Procedure A (Range of temperature): It would give an idea in which part of the year the optimum conditions for germination with regard to temperature prevail within the ecosystem. Soak the seeds in water for 12 to 24 hours.

Sow 100 seeds equidistantly on the filter paper in a petridish. Cover them with another piece of filter paper. Add some distilled water to keep the filter paper moist. Cover the dish. Make 4-5 replicates at each temperature. Observe the seeds daily and record your data. The germinated seeds should be removed from the dishes and a few drops of water should be added daily to keep the seeds moist (or place the filter papers over cotton pads saturated with water in the dish).

	No. of seeds germinated Temperature °C						
Date	10°	15°	20°	25°	30°	35°	40°
Replicate R_1 R_2 R_3 R_4							
Date							
Replicate R_1 R_2 R_3 R_4							

INTERPRETATION AND INFERENCE

(1) Find out the range and optimum temperature of germination. The seeds which do not germinate at any temperature may be (a) Non-viable, (b) Dormant.

(2) The experiment should also be performed with seeds of plants growing in different ecological conditions to demonstrate the variations in optimum temperature and the germination response.

(3) The experiment should be performed after different intervals of seed storage to see if the temperature requirement for germination changes, or is maintained. The gradual loss of viability or release from dormancy can also be inferred from the results.

(4) It may be noted that under optimum temperature conditions, maximum percentage of germination is obtained earlier than under infra or supra-temperature conditions.

(5) Dormancy may be induced in seeds by holding them at unfavourable temperature conditions. This may be shown by placing them subsequently for germination at optimum temperature, where they may fail to germinate.

REFERENCE

Vegis, A. 1963. Climatic control of germination, bud break and dormancy. pp. 265—287. In *Environmental Control of Plant Growth* (Ed. L.T. Evans). Academic Press, New York.

Procedure B (Thermoperiodicity): There are diurnal fluctuations in atmospheric temperature. Seeds of some species become adapted to such conditions for germination process (example: *Eleusine indica*).

Arrange the seeds in petridishes, between two layers of moist filter paper sheets laid over moist cotton pads, and expose them to low and high temperature alternately for different periods.

OBSERVATIONS

Duration in hours at temp.		Percentage germination			
15°C (low)	30°C (high)	R_1	R_2	R_3	R_4
24	0				
20	4				
16	8				
12	12				
8	16				
4	20				
0	24				

INTERPRETATION AND INFERENCES

(1) In what way is this behaviour similar to that of photoperiodism?

(2) Does this germination behaviour help in explaining the distribution of the species?

REFERENCE

Asakawa, S. 1964. The interaction between the thermoperiodic and light sensitive responses in the germination of Ash seeds. *Advancing Frontiers of Plant Sciences* **8:** 1-8.

Procedure C (Photosensitivity): Light plays an important role in the germination process. Seeds of some species are indifferent to light while in many cases it either promotes or inhibits germination.

Arrange seeds in petridishes, and keep them for imbibition in light-proof boxes at the optimum temperature determined by earlier experiments. Maintain one set in dark, while expose the other set to light for different durations daily at fixed time.

Record the actual temperature, light intensity and quality.

After the required lapse of time, to achieve maximum percentage germination, note the results in the following table.

GERMINATION OF SEEDS IN DARKNESS AND LIGHT

Replicates	No. of seeds germinated	
	Darkness	Light duration
1		
2		
3		
4		
5		
Mean ± S.D.		

INTERPRETATION AND INFERENCE

The seeds are photosensitive if they germinate in light, but do not do so in dark. The results under the incandescent and fluorescent light sources may differ, as they vary in the composition of spectral regions.

In what way does the light react?

(*a*) To activate some enzyme system necessary for the process.

(*b*) To destroy a germination inhibitor.

(*c*) To produce a germination promoter.

Does the germination response vary with the age of seeds?

REFERENCE

Toole, H. E., S. B. Hendricks, H. A. Borthwick and V. K. Toole. 1956. Physiology of seed germination. *Ann. Rev. Plant Physiol.* **7**: 299-324.

V. DORMANCY

Dormancy in seeds or buds may be defined as their temporary suspension of growth activity. They do not germinate/sprout under conditions which may be favourable for normal growth. It may be due to:

(1) Enclosing structures, e.g., seed-coat and bud scales restricting oxygen uptake.

(2) Growth inhibiting substances (Wareing 1965).

Dormancy is an ecological adaptation. It delays the occurrence of the species in the ecosystem till suitable environmental conditions for growth are available.

EMBRYO DORMANCY

(*a*) Hasten the after-ripening requirement by keeping the moist seeds at low temperature for varying periods. This process is called "stratification".

(*b*) The enzyme system may be activated by irradiating the dark-imbibed seeds with red light, and then removed to darkness again.

The red light (near about 660 mμ wavelength) may be obtained by monochromatic or interference filters. Light from fluorescent tube filtered through 2 layers of red cellophane paper may also serve the purpose.

(*c*) Vernalization of the seeds at freezing temperatures may also bring about physiological change for germination.

TESTA DORMANCY

The mechanical resistance offered by the seed-coat (or endosperm) may be removed by—

(*a*) Rupturing the enclosing structures by careful hammering.

(*b*) Grinding the seed-coat on a carborundum stone.

(*c*) Placing the seeds in dilute or conc. sulphuric acid for varying periods.

The above treatments will permit water imbibition as well as gaseous exchange necessary for germination.

(*a*) The growth inhibitor may be removed by washing the seeds in running water for several days.

(*b*) Growth-promoting substance may be supplied exogenously.

Prepare Gibberellic acid solution of various concentrations (50, 100, 250, 500 ppm) and soak the seeds with measured quantity of the solution. Chemicals such as thiourea or nitrate may also be used for the purpose.

REFERENCES

Amen, R. D. 1964. The concept of seed dormancy. *Wallerstein Lab. Communication* **27**: 7-19.

Crocker, W. and L. V. Barton. 1953. *Physiology of Seeds*. Chronica Botanica Co., Waltham, Mass.

Evenari, M. 1956. Seed Germination. In *Radiation Biology* (Ed. A. Hollaender) **3**: 519. McGraw-Hill Book Co., New York.

Evenari, M. 1957. The physiological action and biological importance of germination inhibitors. *Symp. Soc. Exptl. Biol.* **11**: 21—43.

Koller, D., A. M. Mayer, A. Poljakoff-Mayber and S. Klein. 1962. Seed germination. *Ann. Rev. Plant Physiol.* **13**: 347—464.

Mayer, A. M. and A. Poljakoff-Mayber. 1963. *The Germination of Seeds*. Pergamon Press, London.

Vegis, A. 1963. (See under exercise 1.25 IVA.)

Wareing, P. F. 1963. The Germination of seeds. pp. 195-227. In *Vistas in Botany* Vol. 3 (Ed. W. B. Turrill). Pergamon Press, London.

Wareing, P. F. 1965. Dormancy in plants. *Sci. Progr.* **53**: 529—37.

VI. REPRODUCTIVE CAPACITY

Salisbury (1942) defines it as the product of average seed output and the fraction represented by the average percentage germination. It can be calculated by the following formula:

$$\text{Reproductive capacity} = \frac{\text{Average seed output} \times \text{Average \% germination}}{100}$$

It is the measure of the intrinsic capacity of a species to increase its number in time, provided the ecizing factors are favourable.

AGGRESSIVE CAPACITY

It can be calculated as below:

$$\text{Aggressive capacity} = \text{Average reproductive capacity} \times \text{Percentage survival of seedlings in nature.}$$

Aggressive capacity denotes the potentiality of a species to colonize and spread in an ecosystem.

REFERENCE

Salisbury, E. J. 1942. *The Reproductive Capacity of Plants.* G. Bell & Sons. Ltd., London.

1.26 Vegetative Propagation

A species perpetuates within the ecosystem by sexual and/or asexual reproduction. The former involves the fusion of male and female gametes, and the production of seed, while in the latter a vegetative part of the plant, e.g., shoot, root or leaf, regenerates a new root system, a new shoot system, or both, by mitotic divisions of cells. Thus in vegetative reproduction, the plants bear the same genetic constitution from generation to generation. Such populations are called *clones.* The plants may produce ecads due to environmental variations, but the morphological differences disappear when the specimens are reared under uniform conditions.

Plants should be observed carefully in the field for methods of their multiplication. Due to biotic operations, such as trampling, grazing etc., the detached fragments of some plants are capable of forming their counter parts. By vegetative propagation, a species may not only survive, but also occupy a larger area in lesser period of time in the ecosystem.

Procedure A (Propagation by stem): Select 8-10 uniform pieces of stem and tag them with labels. Record the number of shoots emerging out per piece at suitable interval of time (weekly). While counting, put iron wire rings in the shoots, so as to differentiate them later from newly formed ones.

OBSERVATIONS

Date		Number of aerial shoots formed Plant number							
		1	2	3	4	5	6	7	8
	Total								

Calculate mean ±S.D.

NOTE: Experiments should be done in the laboratory by planting the stem cuttings in soil or in petridish between moist filter papers.

Determine and record the following:

(1) Optimal temperature.

(2) Time taken for emergence of the root.

(3) Which part of the stem, apical, central or basal or every node—regenerates? Is there any gradient in the regenerative capacity from the base upwards? The stem may be covered with 2-3 cm layer of soil and the rooting at different nodes may be noted.

(4) Make sketches or take photographs.

(5) Cut sections, and examine from where the root initials arise.

(6) Experiments to note the effect of auxins may also be performed.

Procedure B (Propagation by leaf): Cut the leaf into pieces, 5-10 cm long, and insert them in sand or soil. Provide suitable temperature and moisture. Observe if adventitious roots or shoots are formed at the base of the leaf.

Procedure C (Propagation by vegetative bud): In such cases the axillary bud develops into a shoot, and roots are given out at the base.

Place a piece of the stem with a leaf having an axillary bud, in the rooting medium —soil or sand, and provide the necessary environmental condition. Observe and record the growth of the shoot and the root.

Procedure D (Propagation by root): Remove the shoot of the plant in the field. Observe if new shoot develops from the root stock.

Bring the roots to the laboratory and cut 5-6 cm long pieces. Place them below 1-2 cm of soil in a petri-dish or flower pot. Observe the formation of new roots, and the plant.

REFERENCE

Hartmann, H. T. and D. E. Kester. 1959. *Plant Propagation.* Prentice-Hall Inc., Engelwood Cliffs, N. J.

1.27 Seedling Mortality, Growth Rate (Productivity of Different Stages) and Seasonal Periodicity

Generally, all the seeds which germinate in ecosystem, do not develop into mature plants. A large number of them are eliminated on account of death at seedling stage. Thus the population of the species is regulated within the ecosystem.

Procedure A (Seedling mortality): Select 4-5 spots of suitable size in the field when the seedlings first emerge in nature (mostly in the beginning of rainy or winter season). Demarcate the area by fixing pegs. Count the number of seedlings at each spot.

Visit the site every week and record your data in the following table:

Date of observation	Spot No.			
	No. of seedlings survived	No. of new seedlings	No. of seedlings died	Percentage mortality

Plot the results on a graph.

Repeat the experiment in flower beds.

INFERENCE

Assess the reasons for the seedling mortality which may be due to external or internal factors.

External factors:

(1) Unfavourable temperature (too high or too low).

(2) Moisture (Dry condition due to insufficient rains, or waterlogged condition).

(3) Light (Its interception by the foliage of the surrounding plants).

(4) Edaphic factors—low porosity for gaseous exchange of roots, unfavourable pH, saline or calcium rich soils.

(5) Biotic factors—Injury due to trampling or grazing.
Inter -and intra-specific competition for light, moisture and nutrients, etc.

Internal factors:

Effect of seed weight (reserve food material) on the survival of seedlings.

(In *Echinops echinatus*, seed weight varies from 8.0 mg to 18 mg per seed. Seeds weighing 15 mg or above only, develop into mature plants).

Procedure B (Growth rate): A plant develops in an ecosystem by utilizing the matter and energy which are in circulation. As the plant develops, it increases in size, volume and weight (although germinating seed loses in dry weight, due to the utilization of stored food in respiration).

Growth rate may be determined by increase in length, number of leaves, increase in volume, increase in fresh or dry weight. The increase in dry weight is more reliable and often used in growth studies. Dig out five plants of the same age from the field or experimental pots at suitable intervals (every week or fortnight). Wash out their roots carefully and dry them at 60°C for 3 days. The roots and shoots may be dried and weighed separately.

Find out the average $\pm$ S. D. and plot the results in the graph ($\int$-shaped curve may be obtained).

Grow the plants in experimental beds or pots by varying a single environmental factor, e.g. temperature, light, moisture, soil, Ca, N, P, K, Na, etc. and compare the dry weights with those of the control plant.

REFERENCES

Fogg, G. E., 1963. *The Growth of Plants*. Penguin Books Ltd., England.

Whaley, W. G. 1961. Growth as a general process. *Encyl. Pl. Physiol.* **14:** 71-112.

Procedure C (Productivity):

(1) Take out the complete plant from the field.

(2) Estimate the dry weight of various parts, i.e. leaf, stem, roots and seeds, separately.

(3) The exercise has to be carried out at different stages from the seedling stage to maturity.

Record your observations, and represent the data as g of dry matter produced by the species.

1.30 FLOW OF ENERGY

The ecosystem is an open system with input and output of energy. The energy is received from solar radiation. The efficiency with which this input energy can be conserved, determines the status of the ecosystem.

1.31 Radiations (Light and Heat Transfer)

Procedure A (Light):

1. Small booklets, of 15 sheets each of ozalid paper, with pale yellow coated side upward, are prepared.
2. Place the sponge rubber in the petridish, and place a black paper.
3. Place the ozalid paper booklet, with yellow side upward in the centre.
4. Place a black paper on the cover of the petridish from within, and cut out a 2 mm × 2 mm window in the centre.
5. Cover the petridish so that the booklet of ozalid paper is beneath the window.

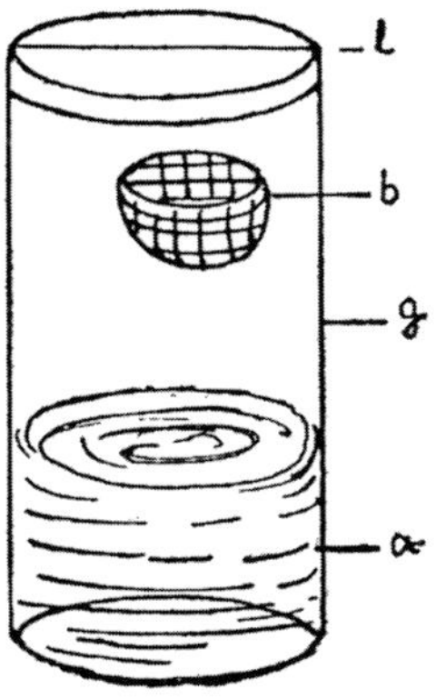

Development jar
(a—ammonium hydroxide, b—wire-net basket, g—jar, l—lid)

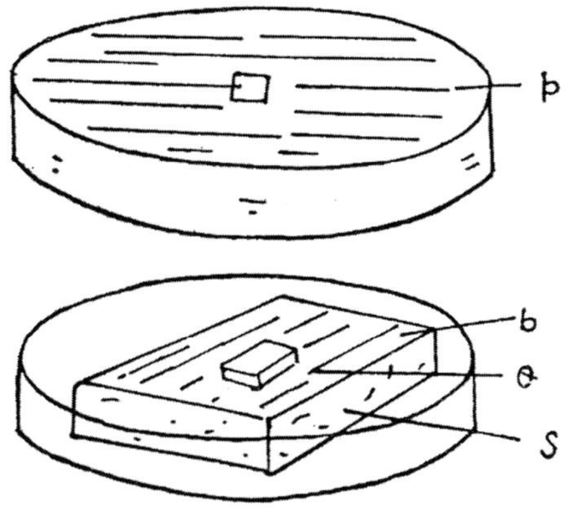

Ozalid paper light measurer
(o—book of ozalid paper, p—black paper, s—sponge)

FIG. 1.31/A

6. All the above operations are done in dim red light.
7. Store such light measurers in light tight containers.
8. Place these light measurers in position, with window facing upwards.
9. Allow this to remain for 2-4 hours and then label the measurer and place in dark containers again.
10. Prepare a development jar as shown in the figure.
11. Place the booklets in a wire net basket in dark and allow it to develop for 10-20 mts.
12. Count the number of layers affected by light.

OBSERVATIONS

Keep at each place at least three measurers and record your data as follows:

Place	Measurer 1	Measurer 2	Measurer 3

INFERENCE AND INTERPRETATION

1. Take these readings inside a grass stand, outside the stand, and at various hours of the day and in various seasons. From this, find out the trend in the change of light intensity at various places.
2. What generalizations can you make out of these observations?
3. Can you suggest any hypothesis for the seasonal and diurnal variations of the light intensity?

REFERENCES

Friend, D.T.C. 1961. A simple method for measuring integrated light values in the field. *Ecology* **42:** 577-580.

Phillips, E.A. 1964. *A Laboratory Book of Field Ecology*. D.C. Heath & Co., Boston.

Procedure B (Light) : Modern light meters are entirely self-contained units. They consist essentially of two parts: a light sensitive cell (photoelectric cell), which generates current when light falls upon it, and a micro-ammeter worked by this current. The current generated in the photoelectric cell is proportional to the light intensity. This current moves a pointer over a scale calibrated in 'lux'. The instrument thus gives a direct reading of illumination at the point where the cell is placed, and in the plane in which the cell lies.

'Lux' is the unit of illumination and is defined as the light flux density on a surface 1 meter distant from a source of 1 standard candle.

Direct exposure of the cell to sunlight will injure the sensitive element. Therefore, in outdoor use, a large piece of white cardboard is placed flat in the direction and location and the reflected light is read. Alternatively a white opaque filter of known factor is placed over the cell. Readings are expressed as a percentage of reading of full sunlight at approximately the same time.

Standardization :

1. Place the cardboard at a distance of 50 cm from the incandescent lamp.
2. Measure the light received directly by the cardboard, by the luxmeter.
3. Measure the light reflected by the cardboard by placing the luxmeter at a distance of 20 cm from the card board.
4. Compute the percentage of light measured by placing the luxmeter at a distance of 20 cm from it.
5. Place the cardboard in the field and measure the light intensity by placing the luxmeter at a distance of 20 cm.

OBSERVATIONS

Record the data as follows:—Percentage of light measured by the luxmeter, by measuring the reflected light from the cardboard. =a.

Places	Reading of luxmeter=r	The intensity of sunlight computed $\frac{r \times 100}{a}$

INFERENCE AND INTERPRETATION

As in exercise 1.31/A.

REFERENCES

Daubenmire, R.F. 1959. *Plant and Environment*. John Wiley & Sons. Inc., N.Y.

Phillips, E.A. 1964. *A Laboratory Book of Field Ecology*. D.C. Heath and Co., Boston.

Trickett, E.S. and L.J. Monsley. 1957. Measurement of radiation. In *Control of the plant Environment* (edited by J.P. Hudson). Butterworths Scientific Publications, London.

Procedure C (Radiant Energy):

CONSTRUCTION OF PYRHELIOMETER

Take a thin silver plate of 2 × 2 cm and get a socket fixed in the middle of the plate. The socket should be such which can hold tightly the bulb of a sensitive centigrade thermometer. The upper surface of the plate is painted black.

1. Weigh the silver plate accurately.
2. Find out the specific heat of the plate.
3. Place the plate and the thermometer in dark before use, and record the temperature.
4. Keep the apparatus in the sun so that the blackened surface faces the sun and is parallel to the ground.

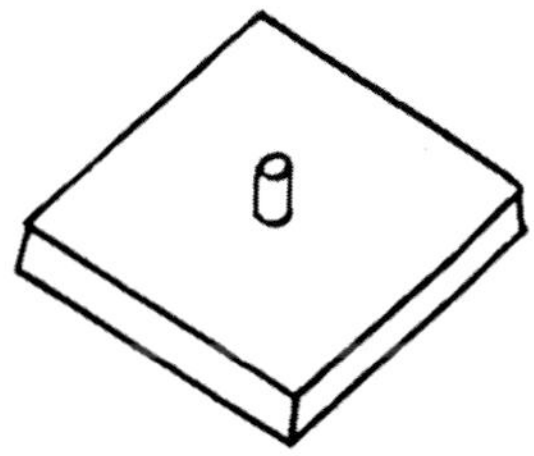

FIG. 1.31/C

5. Allow this apparatus to remain in sun for 5 minutes and record the temperature.
6. Take such readings at different hours of the day and in different months.

OBSERVATIONS

Record your data as follows:—

Weight of the silver plate	x g
Specific heat	s
Time after which the readings are taken	5 minutes
Area of the plate	2 × 2 sq. cm
Constant for a particular plate	$K = \dfrac{x \times 5}{5 \times 2 \times 2}$

S. No.	Date/Time	Initial temperature T	Temperature after 5 mts. exposure T_1	Rise in temperature $T_1-T=t$	Energy in terms of calories $K \times t$
1.	7.00 hrs.				
1.	10.00 hrs.				
3.	13.00 hrs.				
4.	16.00 hrs.				

Average for the day $\frac{\Sigma(K \times t)}{4}$

RESULT

Average energy available is cal/sq. cm/minute.

NOTE

Abbot (1931) and later others, have found out that the average solar radiation is nearly 1.94 cal/sq. cm/minute which varies about 6% during the year, due to the changes in the distances and the angle of the sun from the earth's surface. This value is seen to fluctuate near about 2% during the day.

REFERENCES

Albright, John C. 1939. *Physical Meteorology*. Prentice-Hall Inc., N.Y.

Charney, J. 1945. Radiation. pp. 284-311. In *Handbook of Meteorology* (Edited by F.A. Berry, E. Bollary, and N.R. Berns). McGraw Hill Book Co., N.Y.

1.32 Annual Increment Due to Producers

In computing the annual increment due to producers, care has to be taken to assess the losses due to respiration, consumers and decomposers.

Procedure A (Increase in dry weight):

1. Fix wire netting cages at several places in a field.
2. From three cages take out one monolith of 10 cm square every month.

3. Wash out the soil and find out the oven dry weight of the plant material.
4. Record the dry weights for each month in the following table.
5. Find out the increment or loss in weight of the plant material every month by subtracting the weight of the previous month from the weight of the current month.
6. The algebric sum of all these values for the whole year shall yield the value of the dry matter produced per unit area for a given place per year.

OBSERVATIONS

Months	Dry wt. of plants				Dry wt. of litter				Total	Total gain or loss during the month
	r_1	r_2	r_3	Av.	r_1	r_2	r_3	Av.		
January										
February										
March										
April										
May										
June										
July										
August										
September										
October										
November										
December										

RESULTS

Express the results as g of dry matter produced per unit area per year.

NOTE :—By using wire netting cages we can avoid the macro-consumers, but the result which we get is affected by the losses due to respiration, micro-consumers and decomposers. Thus this method gives only a crude idea of production round the year per unit area of the grassland.

Procedure B (Assimilation): Measurement of the assimilation of a grassland community per square meter per year.

1. Three portable celluloid-covered iron frame work, 1 meter square and 1 to $1\frac{1}{2}$ meter high are prepared. This cabinet has an arrangement for the inlet

of the air from the top and exit from a pipe near the ground. The chamber is air tight. The flow of the air through these chambers is regulated so that from each chamber equal quantity of air is drawn per unit time for sampling.

2. These chambers are fixed in the field, and one is completely covered with a thick black cloth. From under one, the vegetation is completely removed and a celluloid sheet is spread over the ground so that no gaseous exchange can take place even from the soil within the chamber.

3. Arrange the apparatus in the field as shown in Fig. 1.32/B.

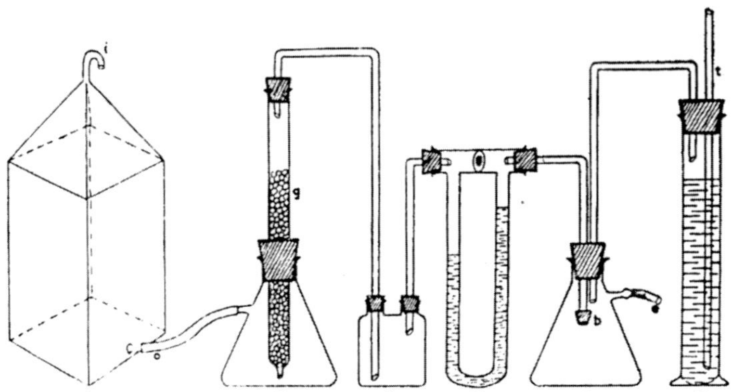

FIG. 1.32/B—Assembly to measure the rate of photosynthesis. A—Chamber, B—Ba $(OH)_2$ tower, C—Trapping chamber, D—Flow meter, E—Filter flask, F—Pressure regulator, b—Bunsen valve, e—To exhaust pump, g—Glass beads, i—Inlet tube, o—Outlet tube, t—By-pass tube

WORKING OF THE APPARATUS

The Ba $(OH)_2$ tower consists of a filter flask containing 200 ml of approximately 0.1N $Ba(OH)_2$ coloured with phenolphthalein. This tower contains a long tube filled with glass beads to ensure a good bubbling of incoming air. A dry safety flask is placed after the absorption tower to catch any accidentally drawn over $Ba(OH)_2$ solution. This leads to a flowmeter which regulates the flow of air through the whole apparatus. This is directly connected to a filter flask attached to the exhaust pump. To the tip of the tube is attached a Bunsen valve so as to check the back flow of the gas. To this filter flask a pressure regulator is inserted to assist in reproducing the experimental conditions and to guard over the excessive suction. The by-pass tube in the pressure regulator should be of 10 or 8 mm bore. It is set near the top of the water at the beginning of the run and is then pressed down until the desired force is exerted on the gas train. It may be necessary to use mercury, if the resistance of the gas train is high. The unequal resistances of different gas trains attached to the same suction pump are adjusted with these regulators.

4. Run this apparatus at various hours of the day for one hour.
5. Take out the $Ba(OH)_2$ solution and titrate against standard HCl (0.1N).
6. Record the readings in the following table.
7. Repeat the experiment at least thrice in a month.

OBSERVATIONS

1 Month	2 Date	3 Time	4 Blank titration	5 Titration of open chamber	6 Titration of covered chamber	7 Titration of the chamber without vegetation	Total Photo-synthesis 5−(4−7)+ 6−(4−7)
January	1st	8-9 A.M.					
		12-1 P.M.					
		4-5 P.M.					
	11th	8-9 A.M.					
		12-1 P.M.					
		4-5 P.M.					
	21st	8-9 A.M.					
		12-1 P.M.					
		4-5 P.M.					
	31st	8-9 A.M.					
		12-1 P.M.					
		4-5 P.M.					

Average per hour for the month $= \Sigma P/12$

Compute the average day length for the month and multiply this average with it.

To get the amount of CO_2 fixed by producer during the month, use the following expression,

$$\text{mg } CO_2 = V_2 \times N \times 22.0$$

where V is the volume, N is the normality.

From the general formula of photosynthesis, we can calculate the amount of energy conserved during the month by the producers.

$$6CO_2+6H_2O+673.2\ Cal - C_6H_{12}O_{12}+6O_2$$

$$\text{i.e. mg } CO_2 \times \frac{673.2}{264} Cal.$$

This all added up for the 12 months will give the total amount of energy fixed by the producers per unit area per year.

RESULT

Express in terms of total amount of energy fixed by the producers per unit area per year.

NOTE:—During recent years many methods for the measurement of rate of photosynthesis have been developed. The apparatus used are very costly and are very technical, hence a very simple method has been described here. Nowadays all the gas analysis is done by Infra-red gas analyser.

REFERENCES

Billings, W. D. 1965. *Plant and Ecosystem.* MacMillan & Co.

Hiniche, A. J. and M. B. Haffman. 1933. The rate of photosynthesis of apple leaves under natural conditions. *Cornell Univ. Agr. Exp. Sta. Bull.* **577**: 1-32; *Science* 77: 55-58, 516-517.

Thomas, M. D., R. H. Hendericks, J. O. Ivie and G. R. Hill. 1943. An installation of large sand-culture beds surmounted by individual air-conditioned green houses. *Plant Physiol.* **18:** 334-371.

Thomas, M. D. and G. R. Hill. 1937. The continuous measurement of photosynthesis, respiration, and transpiration of alfalfa and wheat growing under field conditions. *Plant Physiol.* **12:** 285-307.

Thomas, M. D. and G. R. Hill. 1949. Photosynthesis under field conditions. pp. 19-52. In *Photosynthesis in Plants* edited by J. Francl and W. E. Loomis.

Waugh, J. G. 1937. Precise determination of CO_2 in air. *Ind. & Eng. Chem. Anal. Ed.* **9**: 96-100.

Procedure C (Chlorophyll content): Discs of known size are cut from a number of leaves through a cork borer. Then the known amount (in number) of such discs is crushed in 85% solution of acetone. On centrifuging, the leaf crushes settle to the bottom and the chlorophyll becomes suspended in the acetone solution. The solution is then evaporated to obtain dry chlorophyll and is weighed. This can be expressed as mg chlorophyll/100 sq. cm leaf area.

This procedure has the limitation that it measures apparently acetone extractable substances—all pigments including chlorophyll and some other substances.

Alternatively, optical density of the acetone extract may be recorded at 700 mμ in a photoelectric colorimeter and report the values as such.

1. 33 Food Chain and Rate of Removal of Produce by Consumers

Estimation of the consumers of various trophic levels is very difficult. In this case the aboveground and underground consumers must be estimated separately.

Procedure A:

1. Cages of known size are placed at random at 5 places.
2. The whole container is fumigated with chloroform and all the animals are separated from the vegetation debris.
3. They may be brought to the laboratory for separating the herbivorous and the carnivorous animals.

RESULT

Express the results on per unit area basis.

REFERENCES

Odum, E. P. 1963. *Ecology*. Holt, Rinehart and Winston.

Procedure B (Population and biomass of slightly bigger animals):

1. Trap a few animals from a given area.
2. Mark out the animals with a permanent dye and release them again in the area.
3. Trap out again the animals and make a count of the labelled animals out of the total captured.
4. Find out the average weight of an animal.

OBSERVATIONS

Total number of labelled animals released to the population = a.
Total number of animals captured later on = b
Total number of labelled animals captured later on = c

CALCULATION

$$\frac{\text{Total number of individuals (A)}}{\text{Total no. of labelled individuals released (a)}} = \frac{\text{Total number recaptured (b)}}{\text{labelled in recaptured (c)}}$$

$$\text{Therefore A} = \frac{b \times a}{c}$$

Biomass = Av. wt. of animal × total population.

RESULT

Express the result as the number of individuals (population) or biomass within that area.

Procedure C (Rate of removal):

1. Remove the animals and place a cage of fine wire netting so that it does not permit the entry of even small animals.
2. Place a cage with big holes so that only the grazing animals cannot enter.
3. Periodically determine the amount of producers under the two cages and in the open.
4. The differences in the three cases will give an idea of the rate of removal of produce by various consumers per unit area per unit time.

OBSERVATIONS

Amount of producers in cage 1 = a
Amount of producers in cage 2 = b
Amount of producers outside the cage = c

CALCULATIONS

Amount of produce removed by small animals = a − b

Amount of produce removed by grazers = b − c

RESULTS

Rate of removal of produce by consumers 1 sq. meter/day.

Procedure D (Soil animals):

1. Take a known volume of soil with the help of soil augers.
2. Fumigate the soil with chloroform.
3. Wash out the soil and carefully separate out the animals and count them.

OBSERVATIONS

Record the observations in a tabular form.

RESULT

Results should be given as number per unit volume of the soil.
Alternatively modified 'Berlesi funnel' could be used for this purpose.

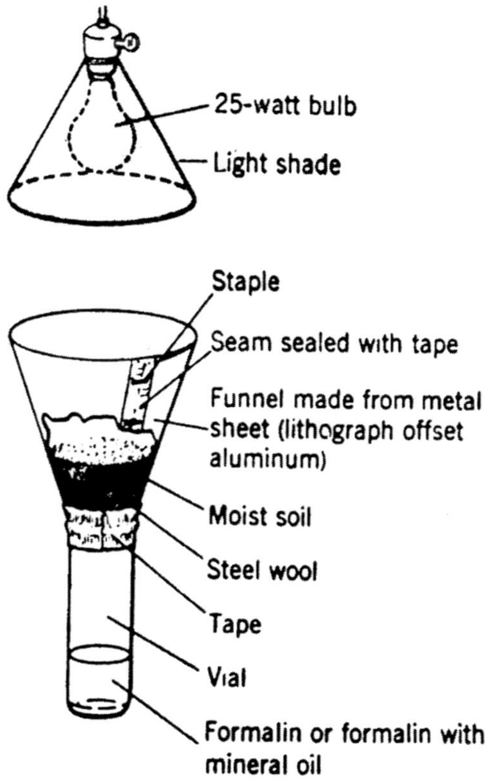

FIG. 1.33/D—Berlesi funnel

1. Take a weighed quantity of soil in the Berlesi funnel and suspend it just above a small container having 5% formalin.
2. Light a small bulb over the funnel and allow it to remain for 7 days. The light and heat of the bulb associated with desiccation of the soil forces the organisms of the soil in the preservative.
3. After seven days count the number of individuals of each species and find out the dry weight. Record your observations in a tabular form.

Amount of animal biomass per g of soil

$$= \frac{\text{Total dry weight of the organisms}}{\text{Dry weight of the soil taken}}$$

REFERENCE

B.S.C.S. Yellow Version Laboratory Manual and Teacher's Guide.

Procedure E (Grazing): The grazing animals affect the grassland ecosystem in a number of ways, namely, by trampling the plant species, by consuming the above-ground parts of the plants, by introducing new species to the plant community and finally by bringing about a number of changes in the soil of the area concerned. The community structure, thus, changes to a great deal due to grazing.

The plant species of a grazed area based on their responses to grazing, may be classified as 'decreasers', 'increasers' and 'invaders'. Decreaser species show maximum density in ungrazed sites, and go on decreasing with increase in grazing intensity. Increasers are those which increase in number when grazing is moderate but with increasing pressure of grazing, their density is also reduced. The increased density of such species under light grazing presumably is the result of the decrease in intensity of competition offered to them by their associates. The invaders are the new additions to the grazed areas. Such plant species are not able to invade the undisturbed grass-land communities, but after grazing they do get the chance to enter and grow.

Select two different sites, of which one should be grazed and the other ungrazed, in a grassland. Make separate lists of the species growing on the two sites, and also find out the density of each plant species. Record the data thus obtained in a table.

INTERPRETATION

Compare the botanical composition of grazed site with that of the ungrazed.

Classify the plant species encountered as decreasers and invaders. In order to recognise the increasers, select one mild grazed area in addition to severely grazed and ungrazed sites. A comparative study of the results obtained in each case will give an idea of the 'increaser' species.

REFERENCE

Curtis, J. T. 1959. *Plant Ecology Workbook: Laboratory, Field and Reference Manual.* Burgess Publishing Co., Minnesota.

Procedure F (Scraping): Besides grazing, a grassland ecosystem is quite often disturbed through scraping, which involves the cutting of the forage species very close to the ground by local people, followed by the removal of herbage from the grassland. The removal of dominant species provides opportunity for invaders to come in, as a result of which botanical composition of the scraped area may change.

Select scraped and unscraped sites in a grassland and examine the botanical composition of each. Also find out the density of the plant species growing on the two sites. Estimate the herbage weight per unit area of the scraped and unscraped sites. Record the data so collected in a table.

INTERPRETATION

1. Name the plant species which are common to both the sites, and those which grow exclusively in scraped or unscraped area.
2. Point out the differences, if any, in the values for density of the species growing on the two sites.
3. Compute the loss in production due to scraping.

Procedure G (Disease): A number of grassland species are attacked by diseases caused by insects or pests. The attacked plants show poor growth as the consequence of which the herbage yield is reduced.

Select the diseased sites in a grassland and estimate the aboveground biomass per unit area. Compare the values so obtained, with the corresponding values for biomass in non-diseased area.

INTERPRETATION

1. Comment upon the type of disease and the symptoms it produces on the plant body.
2. What relation the magnitude of infection bears to the loss in the biomass?

Procedure H (Fire): Fire is among the factors which are destructive to community. Where fire takes place, it burns the aboveground standing crop, and also exerts profound influence upon the disseminules buried in the soil. The subsequent botanical composition of such a community is, therefore, apt to change.

The simplest way to study the effect of fire on a piece of vegetation is through the examination of burnt and unburnt sites in the same grassland. If burnt site is not available, a small section of grassland may be burnt for experimental purposes, and the relevant data may be collected on such a site after a period of 2 or 3 months of firing. Note the number of plant species present on burnt and unburnt sites separately, and also estimate the density of common plant species. Take the weight of the standing crop per unit area of the two sites.

INTERPRETATION

1. Name the species found growing only on the burnt site, and the species which are well represented in the unburnt site, but not occurring in burnt area.
2. Compare the density of the common plant species in the two areas..
3. Indicate, on the basis of your observations, as to which species are sensitive or tolerant to fire.
4. Compute the loss in aboveground biomass of the grassland ecosystem due to burning.

I.34 Energy Release by Decomposers

In the ecosystem, the release of nutrients for recirculation from the dead body of the biomass, is of utmost importance. In soils, there are certain organisms which are responsible for the decomposition of the body and release of materials.

Procedure A (Micro-organisms): Conn's direct microscopic examination of the soil.

1. Place 200 g of soil in each beaker and add the following:—

   (a) No treatment—control

   (b) 1.0g starch

   (c) 1.0g peptone

   (d) 1.0g starch and 1.0g peptone.

2. Mix the substances thoroughly and adjust each sample to 40-50% water holding capacity and allow it to remain so for 1 week.

3. After one week determine the moisture content of the soil.

4. Carefully rule off with India ink 4 squares of 1 sq. cm on each slide.

5. Determine the area of the microscopic field by measuring the diameter with stage micrometer and applying the formula πr^2 to obtain area.

6. Transfer 1g of soil from each beaker into tubes of the fixative solution. Mix well and permit to stand for 1-2 mts.

7. Deliver 0.01 ml of the fixative soil suspension onto the ruled area, spread evenly and permit to dry.

8. After the smears have dried, immerse the slide in 40% acetic acid and then place the slide over a beaker containing water, and the water is kept boiling.

9. Add stain to the slide and allow to remain for 5-6 minutes.

10. Determine the average no. of bacteria in 10 fields, chosen at random, in 1 sq. cm area of each treatment.

OBSERVATIONS

Record the observations as follows:

Soil treatment	Counts/area	Av. of 10 fields	% Moisture	Total count/g dry wt. in million

INTERPRETATION AND INFERENCE

Interpret your data in the light of the materials added to the soils and counts obtained.

REFERENCE

Allen, O. N. 1957. *Experiments in Soil Bacteriology*. Burgess Publishing Co., Minnesota.

Procedure B (Decomposing action):

1. Seal the bottom holes of earthen pots and fill up 5 of them with washed sand, and two with garden soil.
2. Mark off the surface in each case into 3 sectors.
3. Place weighed samples of dead leaves, insects and twigs in each sector, in the pots containing sand and soil.
4. Cover the materials with Petri dish lids.
5. Put the pots in troughs containing water.
6. Weigh the samples after drying at 105°C after fixed intervals.

OBSERVATIONS

Record the dry weight of the samples in the following chart:

Dead object	Changes observed				
	2nd day	I week	II week	III week	IV week

INTERPRETATION AND INFERENCE

1. What substances seem to decompose most rapidly and most slowly on sand and garden soil?
2. What kinds of organisms you observe during the experiment?
3. What general statement you can make out of this experiment?
4. Can you suggest any hypothesis to explain the differences you may observe in sand and garden soil with reference to decomposition rates of various substances?

REFERENCE

B.S.C.S. Green Version—Students Manual.

NOTE:—From the foregoing exercises an energy budget can be prepared showing clearly the amount of energy received from the sun and the amount of it conserved during a year per unit area of the grassland.

1.40 THE WATER CYCLE

The amount of water flowing into the soil and its expenditure from soil during the period considered, characterise the moisture cycle of soil.

Because of the impracticability of the direct measurement of the rates of water flow within the soil the hydrologic balance is expressed as follows:

$$P = E + T + RO + UD + SW + PW + CI$$

Where P=Atmospheric precipitation, E=Evaporation, T=Transpiration, RO=Surface run off, UD=Underground drainage, SW=Change in water content W of the block of soil concerned, PW= Change in plant water content W of the vegetation concerned, CI=Canopy interception.

1.41 Water Input (Precipitation)

Precipitation means the deposition of atmospheric moisture. It may take place either in the solid or liquid form. The various forms of precipitation are:

RAIN—It arises due to the condensation of the atmospheric vapours in the liquid form. It occurs extensively throughout the globe, only exception being the polar areas.

SNOW—The condensation occurs below the freezing temperature and the surface of the earth receives moisture in the solid crystalline form called snow.

SLEET—It is a frozen rain consisting of small solid droplets.

HAIL—It consists of irregularly formed large particles.

DEW—Droplets deposited on cool surfaces.

FROST—Deposition of moisture in solid form on the surfaces which are below the freezing point.

RINE—Deposition of fog particles on the cool surfaces due to the action of wind.

In our region the most important form of precipitation is rainfall. It is measured by a standard Rain-gauge. Total record of the whole year will yield data giving the clear idea of the total water input of the area.

REFERENCE

Foster, E. E. 1949. *Rainfall and Runoff.* The MacMillan Company, New York.

1.42 Return of Water to Atmosphere

I. EVAPORATION

The term evaporation means direct water loss from soil and water surfaces. It is an universal process taking place under all the conditions of nature throughout the world.

The process is governed by the availability of energy at the evaporating surface to meet the latent heat demand. This energy is obtained from the radiations from the sun, sky and clouds. Some of it is, however, supplied by the heat transfers from adjacent air and soil.

The energy distribution at any time depends essentially on the degree of insolation and the reflective power of the surface, the thermal diversity of the soil and air and the vertical temperature gradient. The ability of the air to transfer water vapour away from the surface governs the vertical temperature gradient. This ability in its own turn is dependent upon the vertical gradient of humidity and the effectiveness of turbulent mixing of the lower air layers.

Another important factor governing the evaporation is the availability of water at the effective evaporating surface. In case of land surfaces it depends upon the permeability and the moisture status of the soil. The absolute dryness of the air is not the potent factor in evaporation but the relative dryness of the air and the surface is of importance.

Procedure A: Place the soil of which evaporation is to be measured on the layers of gravel and sand and add water to the gravel layer by subirrigation. Replenish the water from time to time so as to maintain its level constant. Find out the total amount of water added and calculate the rate of evaporation per unit area per unit time.

The results cannot be compared with those under field conditions where no constant supply of water by subirrigation exists.

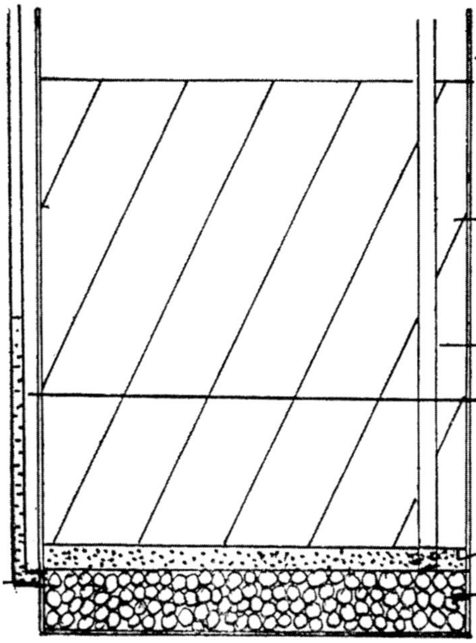

FIG. 1.42/I/A—A convenient apparatus for the measurement of evaporation

Procedure B (Lysimetry): Direct and accurate measurement of evaporation from land surface can be estimated with the lysimeters as given below.

(i) Weighing Lysimeter: Insert an iron drum open at both the ends, one meter long and 60 cm in diameter, in soil. Keeping it in centre, hammer another drum 1.5 m long and 1m in diameter around it. Remove the soil from the portion between the wall of the two drums. Take out the inner drum with the contained soil and attach the concrete holder at its base. Lower the platform balance in the pit and place the lysimeter on it. Record the weight of the whole block periodically and calculate the loss or gain of water due to evaporation or rains as the case may be.

The excess of water collected in the pit during rains should be removed with a water pump.

(ii) Floating Lysimeter: To replace the Weighing of lysimeter by balances it can be installed with a hydraulic system by floating the soil container together with any necessary buoyancy chambers in a suitable liquid held in an outer container, and then detect the changes in the level as the unit rises out of the liquid with loss of weight due to evaporation, or falls back with gain of weight.

The floating lysimeters are relatively insensitive in comparison with those with as ensitive balance. These are liable to errors, due to differential expansion of liquid and pit at different temperatures yet they have the following advantages:

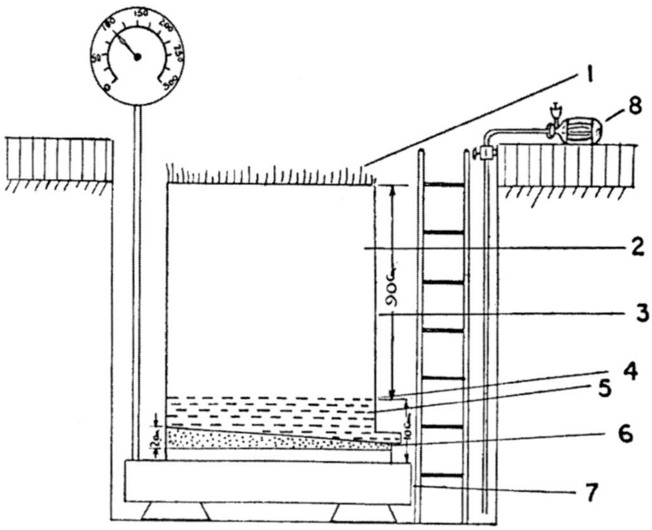

FIG. 1.42/I/B—Weighing lysimeter.

1—Plant cover, 2—Soil, 3—Drum, 4—Perforated concrete base, 5—Pucslate storage space, 6—Concrete slope, 7—Weighing Machine, 8—Pump.

1. Very simple, cheap and require less maintenance.
2. Early adapted to continuous remote recording by using simple type of liquid level recorder with still wells connected to the lysimeter pits by underground pipes.
3. If the inner and outer containers are made leak proof and the annular gap between the two are sealed off by thin polythene sheet possibility of slow zero drift becomes very little.

REFERENCES

King, K. M., C. B. Tanner and V. E. Suomi. 1956. A floating lysimeter and its evaporation recorder. *Tras. Amer. Geophys. Un.* **37:** 738-42.

McIlroy, I. C. and J. Sumner. 1961. A sensitive high capacity balance for continuous automatic weighing in the field. *J. Agric. Eng. Res.* **6:** 252-58.

McMillan, W. D. and R. H. Burgy. 1960. Interception loss from grasses. *J. Geophys. Res.* **65:** 2389-94.

Morris, L. G. 1959. A recording weighing machine for the measurement of evapotranspiration and dewfall. *J. Agric. Eng. Res.* **4:** 161-73.

Popov, O. V. 1959. Lysimeter and hydraulic soil evaporimeters. *Syn. Int. Assoc. Sci. Hydrology* **49:** 26-37.

Pruitt, W. O. and D. E. Angus. 1960. Large weighing lysimeters for measuring evapotranspiration. *Trans. Amer. Soc. Agric. Eng.* **3:** 13-18.

II. TRANSPIRATION

Procedure A: Under field conditions the estimate of transpiration is obtained by using the formula of hydrological cycle, i.e.

$$T = P - E + UD + RO + SW + PW$$

(Refer to water cycle for explanation of symbols)

Lysimetry is the best known method for such a study (Refer to lysimetry, exercise 1.42/IB).

Procedure B (Weighing method): For short period measurements with moist shallow rooted pastures, Pasquill (1949, 50) developed the following technique.

A monolith with the plants is cut from local turf and is placed in a plastic tube. The cavity in the soil is lined from within with another plastic tube in which the monolith (placed in the plastic tube) is inserted. In this way by pulling out the inner tube, the soil core may be removed periodically. At intervals the core is weighed on a weighing machine and returned to its position.

In order to account for the loss in weight due to direct evaporation from the soil surface either the soil surface is coated with some water proof material or by observing the weight loss trend of a similar core without any plant.

Pasquill, F. 1949. Some estimates of the amount and diurnal variation of evaporation from a clay-land pasture in fan spring weather. *Quart. J. Roy. Met. Soc.* **75:** 249-256.

——————.1950. The aerodynamic drag of grassland. *Proc. Roy. Soc.* (London) **A 202:** 143-153.

Procedure C (Cut shoot method): Branches or leaves are detached from the plant and rapidly weighed for fresh weight determination. The material is transferred to original surroundings. The loss in weight is recorded in 2-3 short intervals.

The loss in weight in the given period gives the current rate of water loss due to transpiration.

1.43 Loss of Water Due to Percolation

Infiltration means the absorption of water into the soil from either rainfall or snow-melt. It reduces the runoff and flood flow and replenishes the ground water.

The actual rates of infiltration of the soil in place can be determined by infiltrometers. Essentially these instruments are tight enclosures placed around a small plot of land to which water is applied at a known rate by sprinklers, and from which the run off is caught and measured. Such enclosures may be made of low walls of soil, polythene or metal sheet or wooden planks.

In places where supply of water is frequently an important limiting factor the use of Musgrave's (1935) ring infiltrometer as modified by Marshall and Strick (1950) has been found quite effective.

A ring of 30 cm diameter is inserted into the ground by hammering and jacking. The part of soil disturbed around the ring is pressed back into position. One jet with stop-cock is clamped so that its aperture is 2-3 cm above the soil surface. This is then connected with a graduated water reservoir filled with water and its level is noted. The stop-cock is opened and water is allowed to fill the ring so as to touch the aperture of the jet, which generally takes 5-10 seconds.

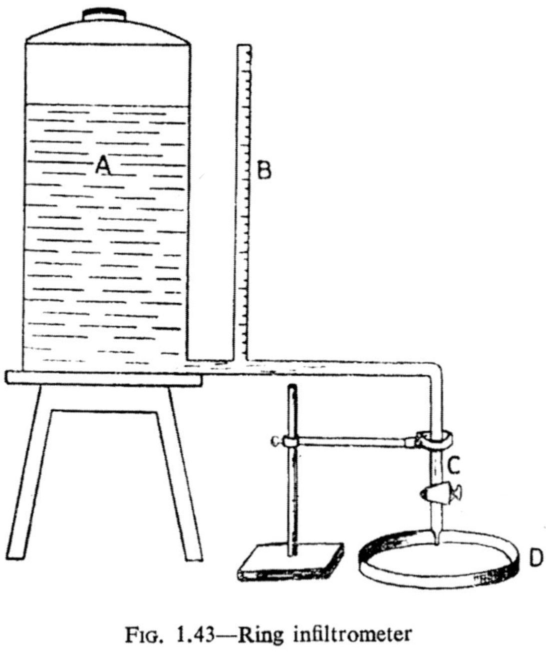

FIG. 1.43—Ring infiltrometer

The opening of the stop-cock is so adjusted that the water level in the ring is maintained at a level. The fall in the level of the reservoir is noted from time to time and then rate of percolation of water is determined.

Marshall and Strick (1950) have introduced a correction factor for the source of error due to lateral penetration of water out side the ring. This is determined by computing the difference between the volumetric water content of the soil of the ring and within one meter of the ring.

The porosity of the soil is the most important factor governing the infiltration rate. It is further governed by vegetal cover, presence of moisture in the soil, etc.

REFERENCES

Marshall, T. J. and G. B. Strick. 1950. The effect of lateral movement of water in soil on infiltration measurements. *Aust. J. Agric. Res.* **1:** 253-65.

Musgrave, G. W. 1935. The infiltration capacity of soils in relation to the control of surface run off and erosion. *J. Amer. Soc. Agron.* **27.**

1.44 Loss of Water Due to Run off

Run off designates all water that flows out and represents the water left over on the surface after evaporation, transpiration and infiltration have taken place.

In case the area is small, the surface run off may be collected along the slopes into a suitable channel through which water may flow out. The volume of flow water may then be determined as in exercise 3.06/C.

For larger catchment areas and valley, measurement of flow in the stream and the rivers is the only way of determining the total run off. In such cases two stations are established in the river or stream, one at the upper altitude and the other at the lower altitude. The volume of flow water may then be determined at both the stations as in exercise 3.06/C. The data obtained at the upper station on being subtracted from that of the lower station gives the surface run off of the area lying between these two stations.

REFERENCE

Foster, E. E. 1949 *Rainfall and Run off.* The Macmillan Company, New York.

1.45 Conservation of Water in The Soil

Procedure A (Water content—gravimetric): Collect about 50-100 grams of soil sample from the site using a tube or auger, at the required depth. Eject the soil into sample cans and close rapidly. Weigh the cans to record fresh weight and dry to give constant weight at 105°C. Note the final dry weight.

$$\text{Moisture percentage} = \frac{\text{Loss of weight on drying}}{\text{Dry weight of soil}} \times 100$$

Procedure B (Water content—volumetric). Excavate the soil by making a hole to desired depth intervals. Collect soil from each interval and store in polythene bags and measure the depth. Weigh the soil and make two or three sub-samples (50 g approx.) and weigh them in sampling cans. Measure the hole with a rule. Continue this process at intervals till the hole reaches the maximum desired depth.

Refill the hole, interval by interval, with sand by pouring sand from a graduated cylinder noting the volume required in each case. Determine the dry weight of the sub-samples gravimetrically.

Calculate the dry weight of the soil in each interval as—

$$\frac{\text{Fresh weight of soil in depth interval} \times \text{Dry weight of sub-samples}}{\text{Fresh weight of sub-samples.}}$$

Calculate the bulk density as—

$$\frac{\text{Dry weight of soil in depth interval}}{\text{Volume of soil in depth interval}}$$

This is expressed as g/cc dry soil.

Alternatively the volume of each interval may be determined by measuring the diameter of the auger hole.

Procedure C (Permanent wilting percentage): Briggs and Shantz (1912) introduced the term 'permanent wilting percentage' or the 'wilting percentage of soils' to designate the moisture content of the soil, expressed as percentage of dry weight, at the time of commencement of permanent wilting in plants growing on the soil. At this moisture level, the water content of the leaves undergoes a permanent reduction to exhibit wilting. The recovery of turgidity is not possible by relieving the plants from transpiration stress, without the addition of water to the soil. Beyond this soil moisture level, further extraction of water may continue at slow rate, but growth is considered as having ceased.

The permanent wilting percentage is a characteristic of the soil and represents the lowest limit of availability of soil water for plant growth. The values of Field capacity and permanent wilting percentage determine the soil moisture availability for plant growth.

Under field condition the soil water content at the root level is progressively determined during the growing season. (Refer Excercise 1.45/A). When the plants seem to wilt permanently as normally evidenced by the failure of the wilted leaves to recover turgidity by morning hours, the soil moisture content on oven dry basis is regarded as the permanent wilting percentage.

Under laboratory conditions Vehimeyer and Hendrickson (1949) suggest to grow sunflower in containers using the given soil. Adequate water is added for the establishment of the plant and uniform root distribution. Plants are then allowed to grow without further addition of water to the soil. When the leaves wilted during the day time, fail to recover turgidity by the next morning, the soil moisture is gravimetrically determined. A series of determinations are made and the average gives the permanent wilting percentage.

If the soil contains less than 0.05% of total soluble salts, the moisture content of the soil at a definite pF value (4.2) obtained by either interpolation or extrapolation, is accepted as the permanent wilting percentage.

Since wilting percentage has definite relations with other soil constants use may be made of some of the following formulae for its determination.

$$\text{Wilting coefficient} = \frac{\text{Moisture equivalent}}{1.84}$$

$$\text{,, \quad ,,} = \frac{\text{Hydroscopic coefficient}}{0.68}$$

$$\text{,, \quad ,,} = \frac{\text{Maximum moisture holding capacity}}{2.90}$$

However, these empirical formulae should be verified through observations and calibrations since the dividing factor is variable subject to texture and conductivity of local soil samples.

LIMITATIONS

Since the permanent wilting percentage as a soil characteristic is determined on the basis of turgor characteristics of leaf tissues, Slatyer (1956) considers it to be an indirect measurement. Plants wilt when the turgor pressure in the leaf decreases to zero and thus permanent wilting percentage determined on the basis of the onset of wilting, using different plants will give variable values.

Under field conditions water extraction often continue long after the plants have wilted particularly in xeromorphic species, which do not visibly wilt at all, water extraction continues even after the growth has ceased. Work of Richards and Weaver (1944) has shown that the range of soil moisture tension at the permanent wilting percentage varies from 10-20 atm., most commonly 15 atm. Determination of Slatyer (1957) suggests that the moisture content of the soil on oven dry basis, which has developed a tension of 15 Atm. (Refer use of tensiometer) gives more reliable value of permanent wilting percentage.

REFERENCES

Briggs, L. J. and H. L. Shantz. 1912. *U.S.D.A. Bur. Plant Indus. Bull.* 230.

Richards, L. A., and L. R. Weaver. 1944. Moisture retention by some irrigated soils as related to soil moisture tension. *J. Agric. Res.* **69:** 215-35.

Slatyer, R. O. 1957. The significance of permanent wilting in studies of plant and soil water relation. *Bot. Rev.* **23:** 586-536.

Veihmeyer, F. J. and A. H. Hendrickson. 1949. Methods of measuring field capacity and permanent wilting percentage of soils. *Soil Sc.* **68:** 75-94.

Procedure D (Moisture equivalent): Briggs and McLane (1907) introduced the expression 'moisture equivalent' to represent the water retaining capacity of the capillary spaces of the soil. They defined it as the amount of water retained by soil when a layer of that soil, 1 cm deep, is centrifuged for 40 minutes in a gravitational field equal to 1,000 times that of gravity. Vehimeyer and others (1924, 27) have modified the definition and may be stated as 'the amount of water expressed as a percentage of oven dry soil, retained by a soil when approximately 30 grams of it are centrifuged for 30 minutes in a gravitational field of 1,000 times that of gravity'.

Moisture equivalent determinations should always be carried out on air dry soils. The soil should be crushed to pass a 2 mm sieve before the determination.

The centrifuge (A) has a special drum head fitted for the determination of moisture equivalent. The drum head (B & C) can accommodate 16 brass cups or boxes with wire net bottoms and sheet metal lids. Each brass cup or box (D) is 5 mm in diameter and 2.5cm deep. The bottoms are curved so as to fit closely against the circumference of the revolving drum head. The centrifuge is fitted with speed controlling rheostat and a speed indicator,

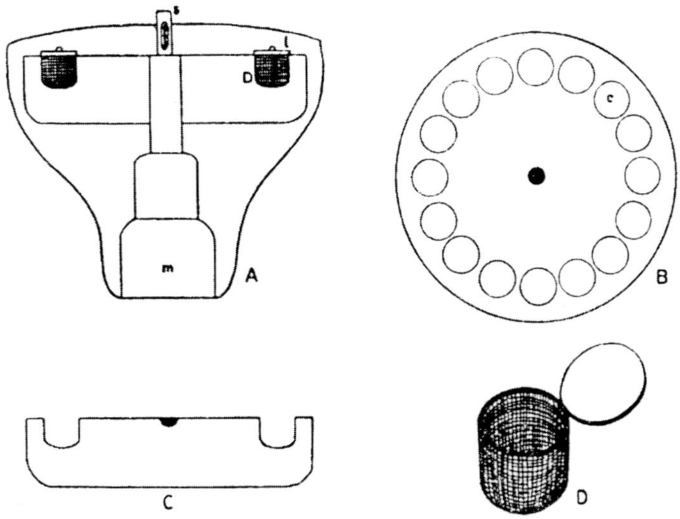

FIG. 1.45/D—Apparatus for determination of moisture equivalent.

The centrifuge should always be run with all the boxes. The cups and lids have numbers and they should be properly placed in corresponding cavities so that weight of each pair placed diametrically opposite is balanced to ensure smooth running of the centrifuge. For the verification of the constancy of the speed control on the centrifuge, a standard soil sample with previously known value should be tried along with the one to be determined. The value should be reproducible within the margin of $\pm$ 1.5%.

Over 30 grams of air dry soil is crushed to pass through a 2 mm sieve. The fine net bottoms of the centrifuge boxes are covered by Whatman No. 2 filter paper and 30 grams of soil ($\pm$ 0.1g) is placed in it. The cup is tapped to level the soil surface. The soil filled cups are then placed on a tray of water to bring the soil moisture level to the maximum water holding capacity. Leaving the cups to stand over night the boxes are allowed to stand for about 30 minutes for the drainage of excess water due to the force of gravity. Necessary protection should be given to prevent evaporation from soil surface. The cups with lids are then placed on the drum head of the centrifuge. The centrifuge is then allowed to revolve and its speed is brought to 2,440 rpm within 3 minutes. The speed is maintained for 30 minutes and then the motor is brought to rest within 3 minutes. The moisture content of this soil is determined gravimetrically (exercise 1.45/A).

Centrifuge boxes can also be prepared by folding a 10 cm square filter paper round a wooden frame. This is placed in a bottomless brass frame.

Coutts (1932) modified the method which requires only 4 grams of soil. A paraffined wooden block with a 1.5 cm diameter hole in the centre is given a shape to fit in the moisture equivalent boxes. About 4g of soil is filled in the hole and the moisture equivalent is determined as usual. The relationship between the method using 30g of soil and that with using only 4g should, however, be checked.

Bouyoucos (1935) described a method where air dry soil is placed on a filter paper in a small Buchner funnel and carefully compacted with soil. The soil is then fully saturated with water for about 24 hours and then the funnel is connected to a good vacuum pump until the whole of the excess water has drained from the top of the soil. After about 15 minutes of further vacuumation, the soil moisture is determined gravimetrically.

LIMITATIONS

Certain types of soils particularly alkali soils on being centrifuged for moisture equivalent determination fail to reach equilibrium as indicated by free water on the surface or by the upper layer of the soil being wetter and softer than the lower layer. This is due to the presence of exchangeable sodium which has become the major portion of the total exchangeable cations.

In some soils separation of clay and bigger particles occur during centrifugation. Presscott and Poole (1934) suggest dilution method in such cases. Soil is mixed with 30, 50 and 70 per cent of pure sand and values are determined. The value of the test sample is determined by extrapolation.

Moisture equivalent may be calculated by using one of the following formulae:

1. ME = Maximum water holding capacity—21 $\times$0.635
2. ME = Hygroscopic coefficient $\times$ 2.71
3. ME = Wilting coefficient $\times$ 1.84

The applicability of the formula is subject to calibration of the correcting factors working with local soil samples.

REFERENCES

Briggs, L J. and J.W. McLane. 1907. The moisture equivalent of soils. *U.S.D.A. Bureau of soils Bull.* **45**, 23pp.

Bouyoucos, G. J. 1935. *Soil Sci.* **40:** 165-170.

Coutts, *J. J. Agric.* **20:** 407-413.

Prescott, J. A. and H. G. Poole. 1934. *J. Agric Sci* **24:** 1-14.

Veihmeyer, F. J., O. W. Israelson and J. P. Conrad. 1924. The moisture equivalent as influenced by the amount of soil used in its determination. *Calif. Agric. Expt. Sta. Techn.* Pap. **16.**

Veihmeyer, F. J., J. Oserkowsky and K. B. Tester. 1928. Some factors affecting the moisture equivalent of soils. *Proc. 1st Int. Congr. Soil. Sci.* **1:** 512-534.

Procedure E (Moistùre tension): Richards and Wadleigh (1952) regarded soil moisture tension as the attraction of the soil for water arising from surface force action at the interphase boundaries in the soil-water system. Thus it has the dimensions of force per unit area and is normally expressed in bars or atmospheres.

INSTRUMENT

Direct field measurements of soil moisture tension up to about 850 mm or in 0-1 atmosphere range can be effectively made with tensiometer. The instrument consists simply of a porous ceramic cup buried in the soil and connected by water filled tubing to a manometer or vacuum gauge on which negative pressure can be indicated. To this is attached a device for filling the water. Diagramatically this is shown in figure 1.45/E. Detailed accounts of these instruments are given by Richards (1949) and Haise (1955).

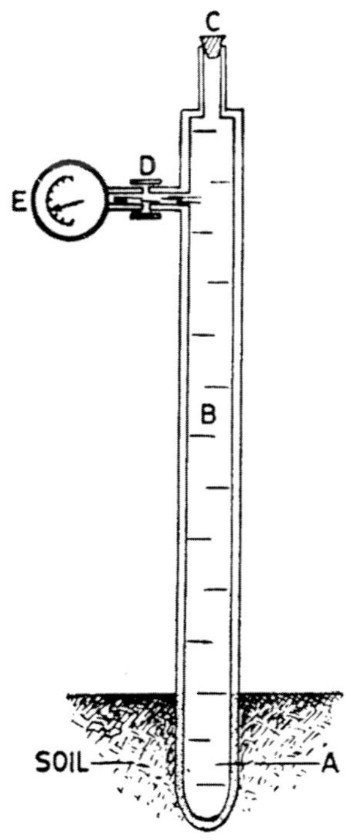

FIG. 1.45/E—Tensiometer (A—porous pot, B—waterfilled tube, C—stopper, D—valve connection, E—vacuum gauge)

Check and eliminate air leaks through tubing connections or in the porous cup by determining the entry valve, which is the air pressure difference required to cause an air leak when the cup is saturated with water. Saturate the cup by wetting under a vacuum, or by the unidirectional water movement through the cup wall. For this carefully fill the cup with a "Squeeze" wash bottle while keeping the outside dry, or by immersing the outside of the cup in water. Make the thoroughly wetted cup empty, immerse in water and connect to a compressed air line. Raise the air pressure until bubbling of air through the cup wall begins. The pressure at this point should exceed one atmosphere.

Fill the instrument with air free distilled water.

Instal by removing a soil core with a tube sample of identical size to the tensiometer tube so as to ensure good soil-cup contact. A separate hole should be drilled to the desired depth for each tensiometer.

The instrument is now operational as soon as soil-cup equilibrium is established.

PRECAUTIONS

In normal installations dissolved air released from the water in the instrument, collects in the air trap while under suction. Whenever this is evident and also following rain or irrigation, the instrument should be serviced and water added.

If the instrument is equipped with a manometer, remove much of the air dissolved in the distilled water, with a syringe through which additional pressure is applied after the air trap has been filled. This forces water through the capillary tubing which connects the porous cup to the manometer. Exclusion of air can be regarded as complete when air bubbles are no longer visible in the water outflow. Refill water with a sqeeze bottle in a vacuum gauge equipped instrument. Remove air from the water with a vacuum pump or as Haise (1955) suggests with bicycle type pump in which the valves have been reversed. Diurnal fluctuations in tensiometer readings frequently occur due to the development of temperature gradients between porous cup and soil. This effect is much smaller in fine than coarse textured soils, and can be minimised if the plastic components are substituted for metals. Readings should be made at the same time each day, preferably early in the morning when the soil temperature gradients are likely to be least steep.

To avoid preferential root growth around the tensiometer cup the location of the tensiometer should be changed periodically.

Because of the restricted tension range in which they operate, the use of this instrument is very limited in naturally arid condition, but can be of considerable value in irrigation studies. It is not useful for small amount of soil as in a pot because it irrigates the soil up to certain extent while measuring the tension.

REFERENCES

Haise, H. R. 1955. How to measure the moisture in the soil *U.S.D.A. Yearbook*: 362-71.

Richards, L. A. 1949. Methods of measuring soil moisture tension, *Soil Sci. 68*: 58-112.

Richards, L. A. and C. H. Wadleigh. 1952. *Soil Water and Plant Growth. Soil Physical Conditions and Plant Growth.* Academic Press, New York. pp. 73-251.

Procedure F (Field capacity): Vehimeyer and Hendrickson (1949) defined field capacity as the amount of water held in the soil after excess water has drained away and the rate of downward movement of water has "materially decreased".

In sand as the large pores are emptied, the permeability decreases more rapidly than gradually, decreasing fine textured soil. Therefore, the transition between supersaturated soil and field capacity is more marked in coarse textured soil.

In soil-water balance studies field capacity has value in representing an upper limit of available water in so far as it is the amount of water which stays long enough in a particular horizon or profile to be of use to plants. This water content is finally reproducible as long as a standardized field procedure is used to determine it.

Lewis (1939) considers it to be the moisutre content of a heavy soil 3 or 4 days after irrigation. In the cases of intermediate to coarse textured well drained soils, it may be regarded as the moisture content a day or two after a rain or after irrigation.

Since field capacity is related to the character of the soil profile laboratory determinations cannot be regarded as exactly equivalent to those obtained in the field. However, columns of soil which are saturated and then allowed to drain while the lower ends are in contact with soil or fine sand will have a moisture content approximating to that of field capacity.

Colman (1947) recommends that the soil be wetted and allowed to drain in sites before sampling for determining field capacity. He found that when a soil was wetted to a depth of 4 ft., soil moisture in the upper 2 ft. was at field capacity after 24 hours. It was recommended by him that the soil be wetted to a considerable depth below the point where samples are to be obtained because shallow wetting results in lower moisture contents than does deeper wetting. Colman (1947) reports that if small blocks of soil are drained on a porous ceramic plate under a tension of 1/3 atmosphere the moisture content will approximately equal to the field capacity of most soils though considerable deviations occur in very coarse and very fine textured soils.

About 100 g of soil is placed in Buchner funnel which is fitted in a filter flask as shown in the figure. A circular filter paper is placed on the perforated disc of the funnel. The soil is then fully saturated with water. Some more water is added and the funnel is covered by a wet cloth and a petridish to check surface drying. The flask is then connected to a water pump to give a tension of 1/3 atmosphere approximately. As the excess water drains down some more water is added at intervals so that about 200 cc of the drained water is collected in the filter flask. When no drop of water is observed to trickle from the funnel in one or two hours' time the moisture content of the soil is determined gravimetrically and is expressed as percentage of oven dry soil at 105°C.

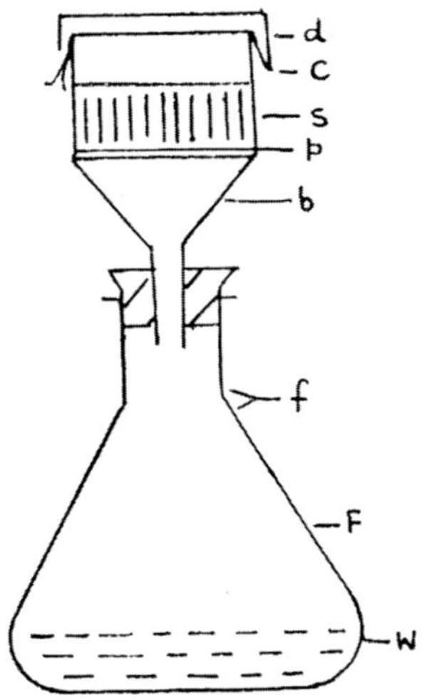

FIG. 1.45/F

(b—Buchner funnel, c—Wet cloth, b—Petridish, f—Filter paper, s—Soil, w—Excess water).

REFERENCES

Lewis, M. R. 1939. Influence of different quantities of moisture on heavy soils on rate of growth of pears. *Plant Physiol.* **10:** 309-323.

Colman, E. E. 1947. A laboratory procedure for determining field capacity of soils. *Soil Sci.* **63:** 277-283.

Veihmeyer, F. J. and A. H. Hendrickson. 1949. Methods of measuring field capacity and permanent wilting percentage of soils. *Soil Sci.* **68:** 75-94.

Procedure G (Sticky point): 'Sticky point' is the moisture content expressed as a percentage of the oven dry soil at which step moist soil just ceases to adhere to external objects. At this state the colloidal components of the soil are saturated with water.

About 20-30g of air dry soil is moistened with small amount of water to make a paste which is moderately wet and sticky. The paste is scrapped from the glass plate and kneaded thoroughly between the fingers, until the paste no longer sticks to the fingers or spatula. At this stage the paste can be clearly cut into pieces without allowing it to stick to the knife. The moisture content of the paste is determined on dry weight basis.

Values for sticky point and moisture equivalent are closely related in most mineral soils with more than 12 to 15% of clay. The value for sticky point is in general, somewhat higher than that for moisture equivalent. The sticky point values for normal clays are probably very close to the moisture equivalents of the same material. However, the organic matter contributes proportionately much more to the sticky point than it does to the moisure equivalent so that the relationship between sticky point and moisture equivalent does not hold for soils rich in organic matter. Sticky point ceases to possess significance in light sandy soils for in these soils the wet sample is not plastic and there is a considerable amount of free interstitial water before exterior adhesion is pronounced. Prescott and Poole (1934) found that as the amount of clay in a soil increases the value for sticky point falls out first but when the amount of colloidal material reaches the equivalent of about 10% of clay it begins to rise in proportion to the colloidal content. The initial decline in the value for the sticky point appears to be associated with the filling up of the pore space in the sandy framework by clay.

REFERENCE

Prescott, J. A. and H. G. Poole. 1934. *J. Agric. Sci.* **24:** 1-14.

Procedure H (Water holding capacity): The maximum water holding capacity of soil is the amount of water taken up by unit weight of dry soil, when immersed in water in a standardized condition. This amount of water is equal to the volume of pores.

Crush air dry soil to pass through 0.5 mm sieve. Place a circular Whatman No. 1 or 44 filter paper on the perforated bottom of the brass box and press it to its position with a split brass ring. Weigh the box and filter paper and record it as W_1.

Fill the brass box with the soil by transferring small portions of soil and tapping the box gently, to ensure its uniform packing. Remove any excess amount by moving one spatula horizontally, touching the upper rim of the box.

FIG. 1.45/H

Place the soil packed box on a Petridish and add water to a depth of 1 cm. Restore this depth after some time and leave overnight. Due to expansion some soil crumbs may fall from the top but cannot affect the determination.

After about 12-16 hours remove the boxes and after surface drying with blotting paper weigh and record as W_2. Place the box in a drying oven at 105°C and record the dry weight after about 24 hours till constant weight is attained, and record as W_3.

Determine the amount of water absorbed by the filter paper by saturating a few filter papers (5 Nos.) with water and by weighing after rolling with a glass rod. Calculate the average amount of water absorbed by one filter paper and record it as W_4.

Calculate the water holding capacity as a percentage of the oven dry soil, as

$$\text{W.H.C.} = \frac{W_2 - W_3 - W_4}{W_3 - W_1} \times 100$$

Knowledge of this value provides a simple means for the determination of desirable moisture level at which the plants can grow well when grown in pots. At the maximum water holding capacity the pF of the soil is zero.

Lack of uniformity in packing of soil and the depth of water surrounding the boxes during soaking may cause inconsistent results.

Procedure I (pF): pF is the logarithm of the capillary potential or the free energy when the latter is expressed in centimeters of water column. It is analogous to the Sorenson's pH scale; the symbol p indicates logarithmic character, while the symbol F denotes free energy. By defining pF as the logarithm of the height in cm of water-column needed to give the suction in question, the logarithm of the free energy difference is measured on a gravity scale.

The entire range of soil moisture tension may be shown on one compact pF scale, although the equivalent water column is almost 1r km at equilibrium with 50 per cent relative humidity (pF6) and about ten times greater at oven dryness (pF7). The moisture equivalent is at pF of about 3.1 the wilting percentage at about 4.2. Water moves along a pF gradient from regions of low pF to regions of higher pF.

The arbitrarily made pF scale ranges from values pF 0 to pF 7.

The significant points on this scale are as follows:—

7.0 Oven-dry soil

6.0 Soil moisture in equilibrium with 50% R. H. (Over 47% H_2SO_4 at 25°C).

5.0 Soil moisture in equilibrium with 94.3% R. H. (Over 10% H_2SO_4).

4.2 Permanent wilting percentage

3.1 Moisture equivalent

2.5 Field capacity

0.0 Soil saturated

It is quite obvious that at all the known soil moisture constants the force of attraction between water and soil particles is the same for all soils, regardless of difference in the percentage of water they contain at this point.

pF values differ depending on whether the equilibrium is reached as a wet soil is being dried out or as a dry soil is being wetted. Only drying values are given in the table.

Because of differences in proportion of different textural components in aggregation and in pore space the pF curves are not identical for different soils, even though all tend to have the same configuration.

1.46 Conservation of Moisture in the Biomass

Out of the total amount of water falling on the ground during precipitation, some amount percolates into the soil and depending on the rate of precipitation some water is lost due to surface run off. The amount of water remaining on the surface ultimately evaporates. The amount of water entering the root layer of the soil is absorbed by the roots, thus increasing the internal water content of the plant.

Excavate the soil before rains and lift some plants with roots and determine the fresh weight. Place them in drying oven at 105°C for 24 hours and determine grams of water present per gram of dry matter. Similarly, after rainfall, lift similar plants from the same site and determine the moisture content of the plants. A few samples should be taken till the surface run off has ceased and the soil is at field capacity.

In this way the amount of water entering into the plant body immediately after precipitation can be measured.

Depending on the density of vegetation estimation of the amount of dry matter per unit land area and accordingly the amount of water which brings changes in the plant water content (△PW in water cycle) can be computed.

NOTE: From the data of the foregoing exercises, a water budget can be prepared, showing clearly the amount of water input and conservation during a year per unit area of grassland.

1.50 CARBON CYCLE

Carbon is absorbed from the atmosphere in the form of carbon dioxide by green plants and is synthesized into complex compounds. The dead plant remains are decomposed under favourable conditions by a variety of micro-organisms which absorb the soluble products as their food. The microbes derive energy from the oxidation of carbon much in the same way as other plants, thus liberating CO_2 in the process. The gas is again absorbed by higher plants during photosynthesis. The carbon cycle is thus seen to be divisible into two phases; one of synthesis of complex compounds and the other of destruction of these compounds. The primary synthesis is based on intake of CO_2 from the atmosphere and while the carbon compounds are elaborated in the biomass at various trophic levels, the use of the food leads to respiratory activity in which process all organisms release CO_2 thus recharging the atmospheric pool of the gas. Taking the biomass of the ecosystem as a whole there is net withdrawal of CO_2 from the atmosphere through the green tissues and net return of CO_2 to it via the respiration of producers and consumers and the activity of decomposers.

Organic matter serves as a source of nutrients and energy in the soil. A soil containing 4% organic matter possesses about 34 metric tons of dry organic substances to the 40.5 hectare furrowslice 910 metric tons. This amount of organic residue possibly carries from 15×10^7 to 18×10^7 kilocalories of potential energy, equivalent in heat value to perhaps 2 to 25 metric tons of anthracite coal. Hence to have an idea of the energy conserved in soil in any ecosystem, it becomes necessary to estimate the amount of organic matter. When organic residues decompose in soil large amount of carbon is liberated in form of carbon dioxide. On the other hand, the nitrogen liberated during decomposition is readily taken up by microorganisms and is therefore, conserved in the soil. This leads to a gradual lowering of carbon/nitrogen ratio of organic residues undergoing decomposition. The decrease in carbon/nitrogen ratio and the rate of evolution of carbon dioxide have been established as an index of rate of decomposition.

1.51 Carbon Dioxide Absorption by the Biomass

The amount of carbon dioxide fixed can be determined as in exercise 1.32/B.

1.52 Carbon Dioxide Loss from the Biomass

The respiratory losses can be found out as in exercise 1.32/B.

1.53 Carbon Dioxide Loss by Soil and Water

Procedure A: Collect sufficient root material of the species to be studied. Spread a weighed quantity (50-100 g) of it over the wire gauze. Bend the wire gauze to form a cylinder (enclosing the bundle of roots). Prepare sufficient number of cylinders and bury them in soil. At monthly or bimonthly intervals remove a few cylinders from soil. Wash the adhering soil. Dry and weigh the root material.

Degree of decomposition can be studied by:

(*a*) Following the loss in weight periodically.

(*b*) Following the decrease in carbon/nitrogen ratio (estimate carbon by Walkley and Blacks' procedure and nitrogen by Kjeldahl method).

NOTE: Decomposition of shoot material can be studied by nylon net technique. For details see exercise 2.53.

Procedure B: Scrap the ground to remove the vegetation. Firmly fix a celluloid chamber or a bell jar on the ground. Connect it to the carbon dioxide trap and proceed as described in Exercise 1.32/B.

Calculate the amount of carbon dioxide by the formula,

$$\text{mg } CO_2 = V \times N \times 22$$

where V = (Blank titration-sample titration, ml)

N = Normality of acid.

Express the rate of evolution of carbon dioxide as mg per hour per unit area for each month. If an estimate of annual carbon dioxide evolution is needed multiply the rate by total hours of the months and add up the values of individual months.

Procedure C:

(1) Place 20 g of soil of each type separately in conical flasks and moisten it with water.

(2) Place a few KOH pellets in the fusion tubes and tie down the same with a thread so as to hang in the conical flask.

(3) Arrange the apparatus as shown in figure 1.53/C.

(4) Mark the initial level at the beginning of the experiment.

(5) After an hour measure the rise in the water level in each case.

(6) Determine the area of the lumen of the tube by measuring its diameter and applying the formula πr^2.

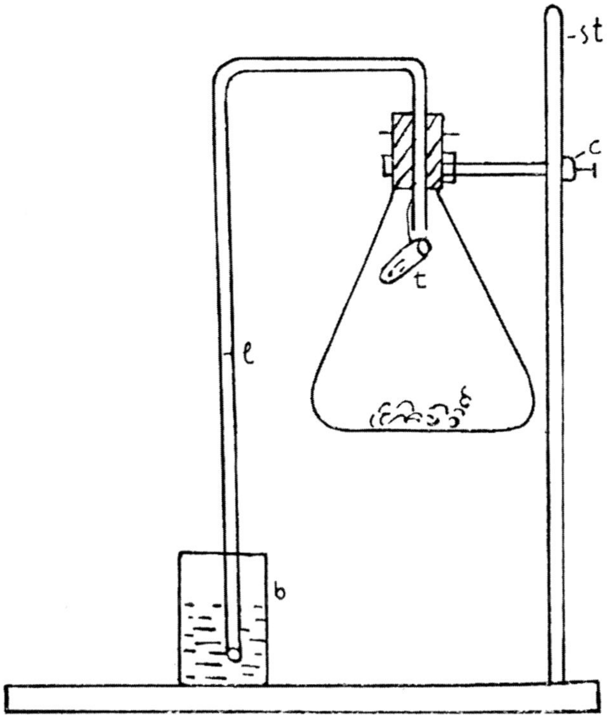

b—beaker, c—clamp, e—eosine solution, s—soil, st—stand, t—small test tube

FIG. 1.53/C

OBSERVATIONS

Tabulate your results as follows:

Type of soil	Rise in level in cm. = l	Diameter of lumen of tube cm = d	Volume of CO_2 evolved $\left(\pi \frac{d}{2}\right)^2 \times l$	Rate of CO_2 evolution

INTERPRETATION AND INFERENCES

(1) In which case the microbial activity is greatest?

(2) What general statement you can make out of this experiment?

(3) Can you suggest any hypothesis to explain the differences you observe in these cases?

1.54 Storage of Carbon as Carbonate and Organic Matter in Soil and Water

Procedure A: Carbonates are decomposed by dilute hydrochloric acid, in vacuum, and the carbon dioxide evolved is absorbed in standard sodium hydroxide. Then barium chloride is added to sodium hydroxide to precipitate carbon dioxide as barium carbonate. Residual alkali is titrated with hydrochloric acid.

The 250 ml flask (B) is fitted with rubber stopper having a separating funnel and a bent tube. The bent tube is connected through a calcium chloride tube with a 1 litre suction flask (D). The flask (D) is connected through a tube (having stop cock) to vacuum pump.

Transfer 0.5—25 g soil to the 250 ml flask. Pipette 50 ml of 0.1 N sodium hydroxide into the 1 litre flask and add a few drops of phenolphthalein indicator. Fix all rubber stoppers tightly and keeping the stopcock F closed, evacuate the apparatus. When evacuation is completed, close stop cock G and pour 50 ml of dilute hydrochloric acid (1 : 20) into the separating funnel. Slowly run down acid into the flask B and allow the reaction to proceed for 15-20 minutes. Connect the gas washing tower (H) to the separating funnel and draw air into the apparatus by partially opening the stop cock F. Shake both flasks at intervals. When vacuum is destroyed, remove flask D from the assembly, and add 10 ml of 15% barium chloride. Titrate the residual sodium hydroxide with standard hydrochloric acid (0.1N) till the pink colour is discharged.

Run a blank determination.

Amount of carbonates expressed as calcium carbonate is calculated as given below:

$$\%\text{calcium carbonate} = \frac{(B-T)}{\text{Wt. of soil}} \times 0.005 \times 100$$

B = ml of 0.1 N hydrochloric acid used in blank.

T = ml of 0.1 N hydrochloric acid used in sample titration.

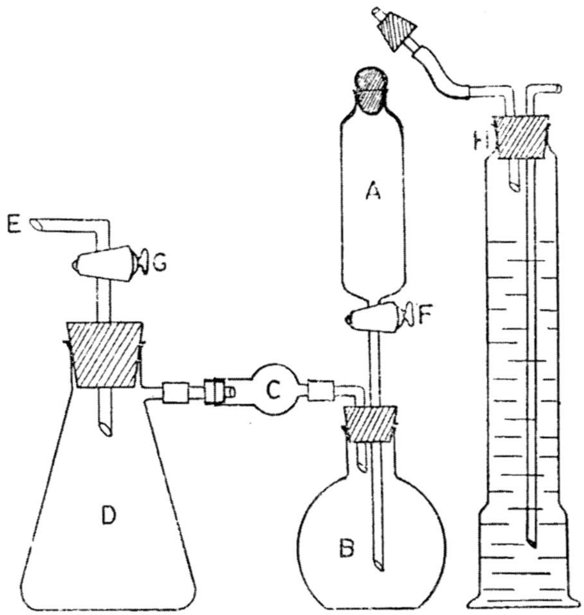

FIG. 1.54/A—Determination of carbonates in soil.

A—Separating funnel, B—Flask, C—$CaCl_2$ tube, D—Suction flask, E—Exhaust pump, F & G—Stop cocks, H—Gas washing tower

In case of soils containing more than 5% carbonates Collin's calcimeter method can be used. For details see Wright (1939).

REFERENCES

Piper, C.S. 1944. *Soil and Plant Analysis*. New York.
Wright, C.H. 1939. *Soil Analysis*. London.

Procedure B: This method is based upon the oxidation of carbon by nascent oxygen, which is liberated from potassium dichromate in presence of sulphuric acid. The reaction is given below:

$$\underset{292.41\,g}{K_2Cr_2O_7} + 4H_2SO_4 = K_2SO_4 + Cr_2(SO_4)_3 \cdot 4H_2O + 3O\ (48g)$$

Thus three atoms of nascent oxygen are liberated from each molecule of potassium dichromate. Hence, 1 ml of normal potassium dichromate is equivalent to 0.008 g of oxygen. 32 g of oxygen combines with 12 g of carbon dioxide. Therefore,

$$32\text{g } O_2 = 12\text{g C}$$

$$\text{or } 0.008\text{g } O_2 = 0.003\text{g C}$$

Hence 1 ml of normal potassium dichromate is equivalent to 3 mg of carbon.

REAGENTS

(*i*) N potassium dichromate solution. Dissolve 49.03g potassium dichromate in water and make up to 1 litre.

(*ii*) Sulphuric acid (at least 96%): 1.25g silver sulphate should be added to each 100 ml of acid.

(*iii*) Phosphoric acid 85%

(*iv*) Diphenylamine indicator: 0.5g diphenylamine is dissolved in a mixture of 100 ml of pure concentrated sulphuric acid and 20 ml distilled water.

(*v*) N ferrous ammonium sulphate solution: Dissolve 393.13g ferrous ammonium sulphate in water and add 15 ml concentrated sulphuric acid to prevent hydrolysis of the ferrous salt. Dilute to 1 litre and standardise against 10.5 ml of N potassium dichromate.

Transfer a weighed amount (not more than 5g) of oven dried soil to a dry 500 ml Erlenmeyer flask and add exactly 10 ml of normal potassium dichromate solution and 20 ml of concentrated sulphuric acid successively. Shake the flask for 2—3 minutes and allow to stand on a sheet of asbestos for 30 minutes. After 30 minutes, dilute the contents of the flask with 200 ml distilled water and add 10 ml phosphoric acid and 1 ml diphenylamine indicator. Titrate the mixture against normal ferrous ammonium sulphate until the colour of the mixture flashes to green. Add an excess of 0.5 ml of dichromate and complete the titration by adding sulphate solution drop by drop until the last traces of blue colour disappear.

Use the following formula to calculate the percentage of carbon.

$$\frac{V_1 - V_2}{W} \times 0.003 \times 100$$ Where V_1 = Volume of N Potassium dichromate (10.5ml)

V_2 = Volume of N ferrous ammonium sulphate in ml.

and W = Weight of soil taken.

Percentage of organic matter is obtained by multiplying the percentage of organic carbon by the factor 1.724. This factor is based upon the assumption that the organic matter of soil contains 58% carbon.

INTERPRETATION

Organic matter is a source of nutrients especially nitrogen, which it releases continually by microbial activity in growing weather. It prevents compaction, increases infiltration, aeration and moisture holding capacity of soil. In general the amount of organic matter is an index of soil fertility.

REFERENCE

Piper, C. S. 1944. *Soil and Plant Analysis*. New York.

NOTE: From the data of the foregoing exercises a carbon as well as an energy budget can be prepared, showing clearly the amounts of carbon fixed and hence the energy received and the amount them conserved during a year per unit area of grassland.

1.60 NITROGEN CYCLE

Nitrogen is essential for growth of any organism because it is the chief constituent of proteins, which are the basis of protoplasm. In any ecosystem air is the reservoir of elemental nitrogen. A small group of free living bacteria and blue green algae in the soil and another bacteria associated with root nodules of members of the family Leguminosae convert large amounts of nitrogen from elemental to complex proteinaceous state. The proteins and allied compounds which mainly constitute the nitrogenous material added to soil, are broken down into proteoses, peptones and finally to amino compounds. This process known as aminization is carried out by a variety of heterotrophic organisms. Amino compounds may either be synthesized into complex cell contents of soil organisms or may be broken into simple products, eventually appearing as ammonia. This process is known as ammonification and is brought about by the same organisms which carry out aminization. Ammonical nitrogen is utilised by ammonifying organisms and many higher plants e.g., rice. The remaining ammonia is readily oxidized into nitrate by certain autotrophic bacteria through a process known as nitrification. Nitrification takes place in two coordinated steps, the first being the conversion of ammonia into nitrous acid, followed by its immediate oxidation into nitrate. Nitrate nitrogen is readily utilised by micro-organisms and most of the higher plants. Under excessive drainage nitrates may be lost in drainage waters, while on the other hand due to poor drainage and aeration, nitrate, nitrate ammonia or even simple amino compounds are reduced and considerable amount of nitrogen escapes from soil in elemental form. This phenomenon is known as denitrification and is facilitated by biotic as well as abiotic means.

Apart from the processes described above, leading to simplification by micro-organisms of complex nitrogenous compounds into simpler forms like ammonia and nitrate, the ecosystem is the seat of great synthetic activity by higher plants. Ammonium and nitrate salts are absorbed by higher plants and synthesized through a number of steps into amino acids and finally into proteins. These proteins are again broken down when dead remains are added to the soil.

In an ecosystem, at any moment, most of the nitrogen is locked up in complex compounds and comparatively very small amount is present as ammonia and nitrate. Hence, to understand the nitrogen economy of any ecosystem, it becomes necessary to estimate the amount of organic nitrogen, ammonia and nitrate in various components of the ecosystem.

1.61 Nitrogen Input by Nitrogen Fixing Organisms

Procedure A (Total nitrogen): Soil nitrogen is mostly present in organic form, and comparatively very small quantities occur as ammonium and nitrate. The Kjeldahl method includes both organic and ammonium forms. The procedure involves digestion of soil sample in concentrated sulphuric acid in presence of a catalyst. Sodium sulphate is added to the digestion mixture to raise the boiling point of sulphuric acid, which should range between 360 to 410°C for proper and quicker digestion. During digestion, ammonia, released by disintegration of complex nitrogenous compounds, is trapped in sulphuric acid to form ammonia sulphate. Later, when the digest is made alkaline and distilled, ammonia is liberated which combines with boric acid to form ammonium borate. Thus the colour of boric acid becomes bluish green due to rise of pH. When this mixture is titrated against sulphuric acid, boric acid is reformed and the pH of the solution is lowered to 4.7 when the indicator turns pink.

REAGENTS

(*i*) Concentrated sulphuric acid

(*ii*) Granulated zinc

(*iii*) Standard N/14 sulphuric acid

(*iv*) Digestion accelerators: 20 g of copper sulphate, 3 g Mercuric oxide and 1 g of selenium powder are mixed by grinding in a mortar, one part of this mixture is thoroughly mixed with 20 parts of anhydrous sodium sulphate and the mixture labelled "Sodium sulphate-plus-catalyst".

(*v*) 40% sodium hydroxide solution.

(*vi*) Mixed indicator solution: 0.5 g brom-cresol green and 0.1 g methyl red is dissolved in 100 ml of 95% ethyl alcohol. This indicator is pink at pH 4.2 or lower and bluish green as the pH rises to pH 4.9 and above.

(*vii*) Boric acid solution: Dissolve 40 g of Boric acid in 1 litre of distilled water containing 5 ml of the mixed indicator. The colour of boric acid is adjusted to faint pink by adding dilute sulphuric acid.

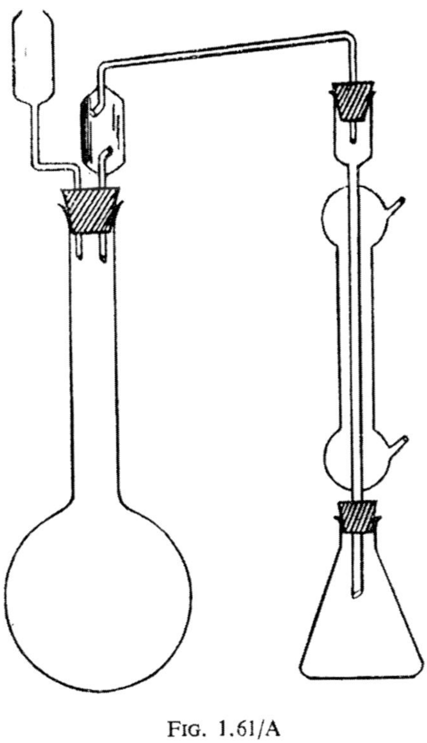

FIG. 1.61/A

Transfer 10 g (1.00 g of peat or 20.0g of sandy soil) soil sample, that has been passed through a 100 mesh sieve, into a 300 ml Kjeldahl flask and moisten it with 25 ml of distilled water. Allow the flask to stand for 30 minutes and then add 20 g sodium sulphate-plus catalyst mixture and 35 ml of concentrated sulphuric acid. Gently swirl the flask to mix the contents and then commence the digestion at low heat. After 10 to 30 minutes, when frothing stops, raise the heat, till the acid boils and condenses approximately one third way up the neck of the flask. Continue the digestion at this heat and rotate the flask at intervals until the organic matter is destroyed. Digestion should be continued for one hour after the digest has become light yellow. When the digestion is completed, stop heating and allow the flask to cool.

When the digest has cooled add 100 ml of water to it, mix well and allow the flask to stand for 3-4 minutes. Then transfer the supernatant liquid into a 800 ml flask, taking care to leave behind as much sand as possible. In the same way carry out 4-5 more washings with 50-60 ml water each time.

Pipette out 25 ml of 4% boric acid into a 500 ml conical flask and add 5 drops of mixed indicator. A glass tube attached to the lower end of the condenser should dip into the boric acid solution. Circulate water through the condenser. Pour 100 ml of 40% sodium hydroxide into the separating funnel keeping the stop cock closed. Open the stopcock of the separating funnel so that sodium hydroxide solution runs into the distillation flask. When only a few drops of sodium hydroxide are left in the separating funnel close the stopcock. Start heating and quickly swirl the flask to mix the contents. Collect 150 ml of distillate and remove the conical flask from the assembly before stopping heat.

Titrate boric acid with N/14 sulphuric acid till the blue colour just disappears and the solution turns faint pink. Carry out a blank determination using all the reagents with 0.2 g sucrose instead of soil. Calculate the percentage of nitrogen by the formula given below.

$$\% N = (T - B) \times N \times \frac{1.4}{S}$$

Where T = sample titration, ml standard acid

B = blank titration, ml standard acid

N = normality of standard acid

S = sample weight, g.

Procedure B (Exchangeable ammonia): The ammonium ion, which is held on the exchange complex, is extracted from the soil by leaching with acidified sodium chloride (sodium being the best replacing ion for ammonium). The estimation of ammonium in the leachate is based upon its reaction with Nessler's reagent to form orange coloured complex, according to the following reactions.

$$\text{Nessler's reagent, } Hg\,I_2 + 2KI = K_2\,Hg\,I_4$$

reaction with ammonia:

$$2K_2\,Hg\,I_4 + 3KOH + NH_3 = Hg_2\,O_2\,(NH_2I) + 7KI + H_2O$$

The percentage transmission of the test solution is read colorimetrically and concentration of ammonium is determined by reference to a calibration curve.

REAGENTS

(*i*) Nessler's reagent: Dissolve 45.5 g mercuric iodide and 35.0g potassium iodide in a small volume of water and transfer the solution into a one litre flask. Add 112g of potassium hydroxide and bring the volume to about 800 ml. When cool make it up to one litre and after allowing the reagent to stand for 3 days, decant the clear supernatant liquid into an amber coloured bottle.

(*ii*) Sodium tartrate: dissolve 100 g in water and dilute to one litre.

(*iii*) Standard ammonium chloride solution: dissolve 1.337 g ammonium chloride in one litre of water containing 1 ml of chloroform. Dilute 20 ml of the stock solution to 500 ml. Each ml of diluted standard solution contains 0.001 ml of NH_3.

(*iv*) Sodium chloride solution: Dissolve 100 g in water and dilute to one litre. Adjust pH of the solution to 2.5 by adding dilute hydrochloric acid.

EXTRACTION OF AMMONIUM

Transfer 100 g freshly collected soil sample into a 500 ml conical flask. Moisture content of the soil is determined separately. Add 200 ml of the acidified sodium chloride solution and shake the flask for 30 minutes. Filter the suspension, through a Buchner funnel fitted with Whatman No.42 filter paper. Collect filtrate in a 500 ml beaker. Rinse the conical flask with 500 ml of sodium chloride solution and pour the rinsings into the Buchner funnel. Leach the soil with additional 200 ml of sodium chloride in four instalments. Transfer the leachate into a 500 ml volumetric flask and make up the volume with acidified sodium chloride solution.

PREPARATION OF CALIBRATION CURVE

Place aliquots of diluted standard (ranging between 3 to 30 ml) in a series of 100 ml volumetric flasks. Add 2 ml of sodium tartrate solution and 50 ml acidified sodium chloride (equal to the aliquot of soil extract used) to each flask. Add water to bring the volume to about 90 ml. Then pipette 5 ml of Nessler reagent and thoroughly mix the contents of the flask. Finally make up the volume, mix and read percentage transmission at 410 mμ, after 25 minutes. Prepare a calibration curve by plotting percentage transmission on a log scale against concentration on a linear scale.

ESTIMATION OF AMMONIUM IN SOIL EXTRACT

Take 50 ml of soil extract in 100 ml volumetric flasks and proceed as in case of standards (delete the addition of acidified sodium chloride). Find the concentration by referring to the calibration curve. Calculate the amount of ammonia as given below:

For 50 ml aliquot:

$$NH_4 \text{ (meq. per 100 g fresh soil)} = 10 \times (\text{meq. } NH_4 \text{ from curve})$$

$$N\text{(part per two million)} = 2800 \times (\text{meq. } NH_4 \text{ from the curve})$$

REFERENCES

Jackson, M. L. 1962. *Soil Chemical Analysis*. Asia Publishing House, Bombay.

Willard, H. H., L. L. Merritt & J. A. Dean. 1958. *Instrumental Methods of Analysis*. D. Van Nostrand Co., N.Y.

Procedure C (Nitrate Nitrogen): Nitrate is extracted from the soil in copper sulphate solution. An aliquot of the extract is evaporated and the dried residue (reaction must be performed in absence of water) is allowed to react, in presence of fuming sulphuric acid, with 2,4-phenol disulphonic acid.

$$C_6H_3\,OH\,(HSO_3)_2 + HNO_3 = C_6H_2OH\,(HSO_3)_2\,NO_2 + H_2\,O$$

The product of nitration which is colourless in acid and yellow in alkaline solution, is made alkaline with ammonium hydroxide to permit colorimetric estimation.

REAGENTS

(*i*) Nitrate extraction solution. Prepare separately 1 N copper sulphate solution

(125 g of $CuSO_4 . 5H_2O$ dissolved in water and diluted to 1 litre) and 0.6% silver sulphate solution (6 gm $AgSO_4$ dissolved in 1 litre of water). Mix 200 ml of 1 N copper sulphate solution and 1 litre of 0.6 per cent silver sulphate solution and dilute to 10 litres. In case of saline soils, containing more than 0.03 per cent chlorine, 2.25 g of powdered silver sulphate for each 1 per cent of chlorine present, should be mixed with the soil sample before extraction.

(*ii*) 2,4-phenol disulphonic acid. Dissolve 25 g pure phenol (crystal white in colour) in 150 ml of concentrated sulphuric acid and then add 75 ml of fuming sulphuric acid. Heat the solution for 2 hours over a water bath and later store in amber coloured bottle.

(*iii*) Standard nitrate solution. Dissolve 0.7221 g pure dry Potassium nitrate in water and dilute to one litre (contains 0.1 g N per ml or 100 ppm). Dilute 50 ml of this solution to 500 ml to get 100 ppm solution.

EXTRACTION OF NITRATE FROM THE SOIL

Transfer 50 g of freshly collected soil sample into a 500 ml extraction bottle and add 250 ml of extraction solution. Determine moisture of soil separately. Fix the stopper tightly and shake the bottle for 10 minutes and then add 0.4 g of calcium hydroxide. After 5 minutes of further shaking add 1 g of magnesium carbonate. These two reagents are used to precipitate the copper and silver and clarify the suspension.* Filter the suspension through a dry filter paper and collect the filtrate in a 500 ml bottle after discarding the first 20 ml filtrate.

PREPARATION OF CALIBRATION CURVE

Place aliquots (2 to 15 ml) of 10 ppm standard nitrate solution in 100 ml silica basin and evaporate to dryness on a waterbath.** Then cool the basin and add 3 ml of phenol disulphonic acid over the dried residue. Rotate the basin to allow the acid to come in contact with all the residue and leave for 10 minutes. Then add 15 ml of water and dissolve the residue by stirring with a glass rod. Make the solution alkaline (yellow coloured) by adding ammonium hydroxide drop by drop and then add an excess of 3 ml. Transfer the solution into a 100 ml volumetric flask and make up to volume. Read the transmission percentage in a colorimeter with 420 mμ filter. Prepare a calibration curve on a semilogarithmic paper, employing the log scale for the transmission percentage against the linear scale for concentration.

*In case of soils yielding strongly coloured extracts add 1g of activated charcoal to the supernatant liquid and shake for 20 minutes. Then add calcium hydroxide and magnesium carbonate as described above, and filter.

**2 and 15 ml aliquots of 10 ppm nitrate solution will be finally diluted to 0.2 and 1.5 ppm.

ESTIMATION OF NITRATE IN EXTRACT

Pipette out 10 ml of soil extract (25 ml if nitrate is expected to be less than 10 ppm) into the basin and evaporate to dryness. Then proceed as in case of the standards. Find the concentration of the test solution from the calibration curve.

CALCULATION

Express the results as parts of nitrate per million parts (ppm) of oven dry soil. Calculate the result as given below:

ppm NO_3 in oven dry soil = ppm NO_3 in test solution $\times$ aliquot dilution $\times$ soil dilution.

$$= \text{ppm } NO_3 \text{ in test solution} \times \frac{\text{ml final colour volume}}{\text{ml aliquot evaporated}} \times \frac{\text{ml extraction solution}}{\text{g oven dry soil extracted}}$$

$$= \text{ppm } NO_3 \text{ in test solution} \times \frac{100}{10} \times \frac{250 + \text{ml } H_2O}{50 - \text{ml } H_2O}$$

[ml H_2O is the volume of moisture present in 50 g fresh soil and is calculated as, 50 g (wt. of fresh soil)—wt. of oven dry soil].

REFERENCE

Jackson, M. L. 1962: *Soil Chemical Analysis*. Asia Publishing House, Bombay.

Procedure D (Nitrification by Bacteria):

(1) Place in each of four beakers 100 g of soil and add following:

a. no treatment—control

b. 0.2g of $(NH_4)_2SO_4$

c. 0.2g of $(NH_4)_2$ SO_4 and 2.5g ground straw

d. 0.2g of $(NH_4)_2$ SO_4; 2.5g ground straw and 0.5g $CaCO_3$.

Mix thoroughly and bring water to 50% field capacity in each case.

(2) Examine the beakers daily and make good the water loss.

(3) Incubate at 25-30°C for 4 weeks.

(4) Estimate quantitatively at the start and the end of the experiment the following:—

a. pH

b. Ammonia

c. Nitrate and nitrite

OBSERVATIONS

Tabulate your results as follows:

Soil treatment	pH		N in mg/100 g of soil			
	Start	End	Ammonical		Nitrite & Nitrate	
			Start	End	Start	End

INFERENCES AND INTERPRETATIONS

(1) In which case do you get maximum nitrification?

(2) What general statement you can make out of this experiment?

(3) Can you suggest any hypothesis to explain the differences you observe in these cases?

REFERENCES

Allen, O. N. 1957. *Experiments in Soil Bacteriology*. Burgess Publishing Co., Minnesota.

Bengston, N. 1924. The determination of ammonia in soil. *Soil Sci.* **18**: 255-278.

Procedure E (Nitrification by leguminous plants):

(1) Place 400 g of soil in each beaker and bring the water to field capacity.

(2) Place seedlings as follows:—

a. No seedlings—control

b. 3 Pea seedlings

c. 3 Gram seedlings

d. 3 Maize seedlings

(3) Examine the beakers daily and make good the water loss.

(4) After three days remove 2 seedlings from each beaker allowing the healthiest to remain in it.

(5) Allow the experiment to continue for 4 weeks.

(6) Estimate quantitatively at the start and end of the experiment the total nitrogen of the soil.

OBSERVATIONS

Record your results as follows:—

Experiment 1	Total N_2 at the start in mg/100 g of soil 2	Total N_2 at the end in mg/100 g of soil 3	Gain or loss of nitrogen (3−2) 4

INFERENCES AND INTERPRETATIONS

(1) In which case you get maximum nitrification and maximum loss of nitrogen?

(2) What general statement can you make out of this experiment?

(3) Can you suggest any hypothesis to explain the differences you observe in these cases?

1.62 Nitrogen Loss by Denitrifying Organisms

(1) Inoculate Durham tubes as follows:—

a. no treatment—control

b. 1 ml soil suspension

c. 1 ml manure suspension

d. 1 ml mud of bog suspension.

(2) Incubate the tubes at 30°C for 2 weeks.

(3) Estimate the following at the beginning and the end of the experiment:

a. pH

b. Ammonical nitrogen

c. Nitrite and nitrate nitrogen.

OBSERVATIONS

Record the data in the following manner:—

Treatments	pH		Ammonical N		Nitrite & Nitrate N	
	Beginning	End	Beginning	End	Beginning	End

INFERENCES AND INTERPRETATIONS

(1) In which case do you get maximum denitrification?

(2) What general statement can you make from this experiment?

(3) Can you suggest any hypothesis to explain the differences you observe in these cases?

REFERENCES

Allen, O. N. 1957 *Experiment in Soil Bacteriology.* Burgess Publishing Co., Minnesota.

Bengston, N. 1924. The determination of ammonia in soil. *Soil Sci.* **18:** 255-278.

1.63 Uptake and Storage of Nitrogen in Plants

Procedure A (Total nitrogen of plant): This method is essentially similar to the macro-Kjeldahl procedure. The only difference being the distillation by steam heating in a specially designed apparatus (micro-Kjeldahl distillation unit).

REAGENTS

(*i*) Standard hydrochloric acid solution of N/28 strength. Dilute 31.8 ml concentrated hydrochloric acid to 10 litres with water and standardise against standard borax or sodium carbonate solution.

(*ii*) Catalyst. 1 g copper sulphate, 8 g potassium sulphate and 1 g selenium, dioxide are powdered separately and then thoroughly mixed together.

(*iii*) Mixed indicator. Mix 6 ml methyl red solution (0.16% in 95% alcohol) and 12 ml brom-cresol green (0.04% in water). Add 6 ml of 95% alcohol to the mixture.

(*iv*) Concentrated sulphuric acid (N-free).

(*v*) Boric acid. Dissolve 20 g of boric acid in water and dilute to 1 litre.

(*vi*) Sodium hydroxide. Dissolve 500 g in 1 litre of water and allow to stand for a few days. Siphon the clear supernatant in another bottle and label as 40% sodium hydroxide.

DIGESTION

Wrap a 250 mg dried plant tissue sample (ground to pass 40 mesh sieve) in a small piece of qualitative filter paper and drop the package into a 50 ml digestion flask. Add 5 ml of concentrated sulphuric acid and approximately 0.1 g of catalyst mixture. Digestion is performed at low heat until frothing stops and fumes of sulphuric acid are freely evolved. After 5-10 minutes increase heat so that the acid boils and condenses one third way up the neck of the flask. Continue digestion for at least 30 minutes after the digest has become apple green. On completion of digestion, cool the flask and add 20 ml of water. Again cool the flask and transfer the contents into a 50 ml volumetric flask and make up the volume.

DISTILLATION

Distil in a Markham apparatus. Set up the assembly as shown in figure. Boil distilled water in the flask (A) and keep the clips (B) and (C) closed and open respectively. When steam passes through the steam jacket (D), pipette out 10 ml aliquot of digest into the distillation chamber (E) through the funnel (F) and wash the latter (F) twice with 1 ml water each time. Replace the ground glass stopper and add 5 ml of 40 per cent sodium hydroxide in the funnel (F). Keep the lower end of the condenser dipped into 5 ml of 2 per cent boric acid contained in a 50 ml conical flask. When steam issues freely through the tube (G), close the clip (C) and lift the ground glass stopper to allow sodium hydroxide to run into the digest. Immediately replace the stopper and continue distillation until 30 ml distillate has been collected. After a few ml liquid has distilled over the end of the condenser should be raised above the level of boric acid. When distillation is complete, stop heating

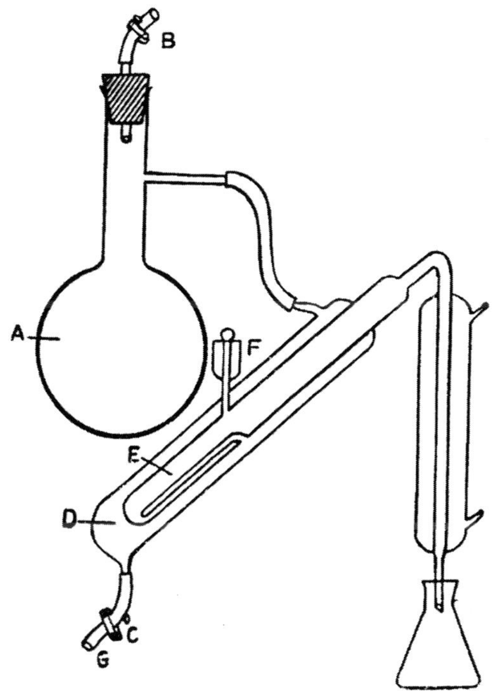

FIG. 1.63/A. Markham steam distillation assembly.

so that the liquid in distillation chamber (E) is automatically sucked into the jacket (D). Add a few ml of water through the funnel and replace the stopper. Allow the liquid in the jacket to run to waste by opening the clip (C). Now the apparatus is ready for the next distillation.

CALCULATIONS

Calculate the result from the following formula :

$$\%N = (T\text{-}B) \times 10 \times N \times \frac{1.4}{S}$$

where T = sample titration, ml

B = blank titration, ml

N = normality of hydrochloric acid

S = weight of plant material, g

The difference T-B is multiplied by 10 because only 10 out of 100 ml digest was distilled.

Procedure B (Total nitrogen including nitrates): Nitrates are included by the addition of salicylic acid which becomes nitrated and the nitro-compound breaks down to ammonia during digestion.

REAGENTS

(*i*) Salicylic-sulphuric acid mixture. Dissolve 5 g salicylic acid in 100 ml concentrated sulphuric acid.

(*ii*) Sodium thiosulphate.

After transferring the weighed plant sample to the Kjeldahl flask add 5 ml of salicylic-sulphuric acid mixture. Rotate the flask to mix and allow to stand for 30 minutes. Add approximately 0.3 g sodium thiosulphate and heat gently until fumes appear. Then add catalyst and sulphuric acid and proceed as in the previous experiment.

Procedure C (Nitrate nitrogen): This method also employs phenoldisulphonic acid-nitrate reaction.

REAGENTS

(*i*) Copper sulphate solution. Dissolve 5 g of $CuSO_4 . 5H_2O$ in water and dilute to 1 litre.

(*ii*) Silver sulphate solution. Dissolve 3.5 g silver sulphate in warm water and dilute to 1 litre.

(*iii*) Calcium hydroxide—magnesium carbonate mixture. Powder and mix 10 g of calcium hydroxide and 20 g of magnesium carbonate in a mortar.

(*iv*) Phenol-disulphonic acid (see exercise 1.61C).

(*v*) Sodium phosphate solution. Dissolve 138 g of NaH_2PO_4 in 500 ml water, add sodium hydroxide solution to bring the pH to 6.5 and dilute to 1 litre.

(*vi*) Standard potassium nitrate solution. Dissolve 0.5055 g of pure potassium nitrate in water and dilute exactly to 1 litre.

(*vii*) Ammonium hydroxide. Mix equal volumes of ammonium hydroxide and water.

Transfer 100 mg of finely powdered plant material into a 25 ml beaker and add 9 ml of silver sulphate solution. Swirl the beaker and quickly add 1 ml of sodium phosphate solution in order to precipitate the excess of silver. Allow to stand for 2 hours and then filter through Whatman No. 42 filter paper. Place 2 ml aliquot of the extract into a 15 ml centrifuge tube; add 2 ml of copper sulphate solution and 2 ml of water. Add 0.5 g of calcium hydroxide-magnesium carbonate mixture and allow to stand for 1 hour before centrifuging. Evaporate to dryness (on a water bath) 2 ml of supernatant liquid in a 25 ml beaker. Allow to cool and rapidly deliver 1 ml of phenol-disulphonic acid over the residue.

Rotate the beaker and after standing for 10 minutes add 10 ml of water. Stir with a glass rod to bring the residue in solution. Make distinctly alkaline with ammonium hydroxide. Transfer into a 25 ml volumetric flask and dilute to volume.

PREPARATION OF CALIBRATION CURVE

Pipette into centrifuge tubes varying amounts (0, 1, 2 and 3 ml) of standard nitrate solution and add sufficient water to make the volume to 4 ml in each tube. Then add 2 ml of copper sulphate and 0.5 g of calcium hydroxide-magnesium carbonate mixture. Proceed further as in case of unknown solution.

Measure the percentage transmission of standards at 420 mμ. Prepare calibration curve on a semilogarithmic paper using the log scale for percentage transmission.

CALCULATION

Find the concentration of the unknown solution from the calibration curve. 1, 2 and 3 ml of the standards represent 0.07, 0.14 and 0.21 mg of nitrate. Calculate the percentage of nitrate in the plant material by the formula given below:

$$\text{Percent nitrate in the plant material} = \frac{\text{mg nitrate (from the curve)}}{1000} \times \frac{\text{extraction solution ml}}{\text{aliquot ml}} \times \frac{100}{\text{wt. of material g}}$$

According to the procedure outlined above

$$\text{Percent nitrate in plant material} = \frac{\text{mg nitrate (from the curve)}}{1000} \times \frac{10}{2} \times \frac{100}{0.1}$$

$$= \text{mg nitrate} \times 5 \text{ (from the curve).}$$

REFERENCE

Peach, K. and M. V. Tracey. 1956. *Modern Methods of Plant Analysis*. Volume I, Springer-Verlag. Berlin.

NOTE:— From the data of the foregoing exercises a nitrogen budget can be prepared, showing clearly the amount of nitrogen fixed and conserved during a year per unit area of grassland.

1.70 CYCLING OF Ca, Mg, P, K

Available minerals are extracted from the soil by absorption through roots into the plant body. They remain locked there, participating in the metabolism until taken up by other organisms through plant food. On death and decomposition of the bodies of the organisms these minerals are returned to the soil again in the available form.

1.71 Uptake and Storage in Plants

REAGENT

Dilute hydrochloric acid: Dilute concentrated acid with equal amount of water.

Procedure A (Dry ashing): Weigh 5 g oven dried powdered sample into a silica basin (previously ignited to constant weight). Char the material over a Bunsen burner and transfer the basin to a muffle furnace. Maintain the muffle furnace at about 300°C, till the material ceases to glow and then raise the temparature to 450—500°C. Continue ignition till a white ash is obtained. Cool the basin in a desiccator and weigh. Find out the weight of ash and report as the percentage of original oven dried material.

Carefully add about 40 ml of dilute hydrochloric acid to the basin (avoid spattering) and digest over a water bath for an hour. Add 20 ml of hydrochloric acid and 50 ml of water; warm on a water bath and filter through a Whatman No. 42 filter paper into a 250 ml volumetric flask. Transfer the residue to the filter paper and wash several times with hot dilute hydrochloric acid. If necessary ignite the filter paper, weigh the residue and report this as silica (SiO_2). Make up the volume of the filtrate to 250 ml. Aliquots of this ash solution are used for determination of major elements. Alternatively ash solution can be prepared by the Procedure B.

Procedure B (Wet ashing): Transfer 5 g oven dried sample into a 300 ml Kjeldahl flask and add successively 4 ml perchloric acid, sufficient nitric acid (about 35 ml) and 5 ml sulphuric acid. Swirl the flask to mix and heat gently at low heat for 5 minutes, till dense brown fumes appear. Remove the flask from the heater after 5 minutes. Replace the flask on the heater and continue digestion at low heat, till white fumes appear. Continue digestion at low heat for 5-10 minutes after the appearance of white fumes. Finally digest at high heat for 1-2 minutes. It is important to control the rate of digestion and if properly done the whole operation will take 90-100 minutes. On completion of digestion the liquid becomes colourless. If some carbon particles are still left, add 2 ml of nitric acid and digest to fuming. Cool and dilute with 100 ml of water. Transfer into a 250 ml volumetric flask and make up the volume. This solution is to be used for analysis of the major elements.

Procedure C (Estimation of calcium):

REAGENTS

(*i*) Saturated ammonium oxalate solution. Dissolve approximately 40 g in 1 litre of water.

(*ii*) Methyl red indicator. 0.125% alcoholic solution.

(*iii*) Ammonium hydroxide. Dilute 1 part concentrated ammonium hydroxide with 4 parts of water.

(*iv*) Dilute sulphuric acid. Dilute one part concentrated acid with four parts of water.

(*v*) 0.05 N potassium permanganate (see exercise 1.72/B).

Pipette 50-100 ml aliquot of ash solution into a 250 ml beaker. Heat to boiling and add 10 ml of hot ammonium oxalate solution and 1 drop of methyl red indicator. Add ammonium hydroxide drop by drop till solution becomes pink and boil for 2 minutes. Allow to cool and adjust the colour of the solution to faint pink by addition of ammonium hydroxide. Allow to stand for 4-5 hours and filter. Wash the precipitate 5-6 times with water. Save filtrate for estimation of magnesium. Pierce the filter paper by means of a glass rod and wash the precipitate into a beaker (in which precipitation was made). Wash the filter paper alternately with hot water and hot dilute sulphuric acid. Finally add 20 ml of dilute sulphuric acid, heat to 80-85°C and titrate with 0.05 N potassium permanganate, till pink colour just persists. Just before end point add the filter paper to the solution and complete the titration.

CALCULATIONS

1 ml of 0.05 N potassium permanganate = 1 mg calcium.

$$\% \text{ calcium} = (\text{wt. of calcium in aliquot g}) \times (\text{aliquot factor}) \times \left(\frac{100}{\text{wt. of sample, g}}\right).$$

Procedure D (Magnesium):

REAGENTS

(*i*) Concentrated hydrochloric acid.

(*ii*) 10% sodium citrate. Dissolve 50 g in water and dilute to 500 ml.

(*iii*) 10% ammonium phosphate $[(NH_4)_2\ HPO_4)]$. Dissolve 50 g in water and dilute to 500 ml.

(*iv*) Concentrated, (1+4) and (1+10) ammonium hydroxide.

Concentrate the filtrate from calcium determination to about 100 ml and add 5 ml of concentrated hydrochloric acid. Heat to boiling and add 5 ml of 10% sodium citrate and 10 ml of 10% ammonium phosphate. Make slightly alkaline by adding dilute ammonium hydroxide (1+4) with constant stirring. When the solution becomes turbid, add 25 ml of concentrated ammonium hydroxide and stir rapidly for a few minutes. Allow to stand overnight and filter through a Whatman No. 42 filter paper. Wash the precipitate with dilute ammonium hydroxide (1+10) until free from chlorides. Ignite the filter paper with precipitate, cool in a desiccator and weigh.

CALCULATIONS

Find the amount of magnesium in the aliquot by multiplying the weight of the the pricipitate by 0.218.

$$\%\ \text{magnesium} = (\text{wt. of magnesium in aliquot, g}) \times (\text{aliquot factor}) \times \left(\frac{100}{\text{wt. of sample, g}}\right).$$

Procedure E (Potassium):

REAGENTS

(*i*) 0.02N oxalic acid. Dissolve exactly 1.2604 g in 200 ml water. Add 10 ml concentrated sulphuric acid and dilute to 1 litre.

(*ii*) 0.02 N potassium permanganate. Prepare solution containing 0.64 g potassium permanganate per litre (see Excercise 1.72·1B).

(*iii*) 10% hydrochloric acid. Dilute 10 ml of concentrated acid to 100 ml.

(*iv*) Concentrated nitric acid.

(*v*) Acetone.

(*vi*) 1.4% silver nitrate. Dissolve 1.4 g in water and dilute to 100 ml.

(*vii*) 25% sodium cobaltinitrite. Dissolve 25 g in water and dilute to 100 ml. Store in a refrigerator.

(*viii*) Glacial acetic acid.

(*ix*) 2N sulphuric acid.

Evaporate to dryness an aliquot (5-10ml) of ash solution with 2 ml nitric acid in a 25 ml beaker. When cool take up in 0.5 ml of water, warm and transfer to a centrifuge tube. Rinse the beaker thrice with 0.5 ml warm water into the tube. Add successively a drop of glacial acetic acid, 0.5 ml 1.4% silver nitrate, 1 ml 80% acetone, stirring after each addition. Cool the tube to 2-4°C and add 1 ml of 25% sodium cobaltinitrite (previously cooled to 2-4°C), rinsing the glass rod with last few drops. Allow to stand at 2-4°C for 2 hours.

Centrifuge for a few minutes and discard the supernatant. Allow the tube to drain by inverting over a filter paper. Slowly add 3 ml of 50% acetone along the sides of the tube and gently swirl. Centrifuge, discard the acetone and allow to drain. Wash the precipitate thrice with 80% acetone, stirring each time with a glass rod. Finally suspend the precipitate in 3 ml of pure acetone, rinse the rod and centrifuge. Allow to drain. When the precipitate becomes dry, suspend in 3 ml conical flask. Wash the tube with water, then with 5 ml of 0.02 N potassium permanganate and finally thrice with water, adding the washings to the flask. Pipette 15 ml of 0.02 N potassium permanganate into the flask. Heat the contents of flask to boiling and then rapidly cool under running water. Add 20 ml of 0.02 N oxalic acid. Titrate with 0.02 N potassium permanganate.

CALCULATIONS

Amount of potassium (mg) in the aliquot is calculated as follows:

Total amount of 0.02 N $KMnO_4$ (ml)—Oxalic acid (ml) $\times$ 0.1086 (i.e. 20+titration value)

Hence % of potassium

$$= \text{Amount of potassium in the aliquot (g)} \times \text{aliquot factor} \times \frac{100}{\text{wt. of sample, (g)}}.$$

Procedure F (Phosphorus): Plant material is ashed in presence of magnesium nitrate to prevent the loss of the phosphorus. The colorimetric estimation of phosphorus in ash solution is based upon the formation of blue complex between phosphate and molybdic acid in presence of a reducing agent.

REAGENTS

(*i*) 0.1 N Magnesium nitrate. Dissolve 3.71 g in 500 ml water.

(*ii*) 10 N sulphuric acid. Dilute 300 ml concentrated acid with 867 ml water.

(*iii*) Ammonium molybdate solution. Dissolve 8.3 g in 100 ml warm water.

(*iv*) Stannous chloride solution. Dissolve 0.2 g in 0.5 ml concentrated hydrochloric acid and dilute to 100 ml. Use freshly prepared solution.

(*v*) Standard phosphate solution. For stock solution dissolve 1.097 g pure potassium hydrogen phosphate (KH_2PO_4) in water and dilute to 250 ml. Dilute 2.5 ml of stock solution to 250 ml to prepare working standard.

ASHING

Mix thoroughly 50-200 mg sample with 5 ml 0.1 N magnesium nitrate in a silica basin. Evaporate to dryness and then ignite at 500°C for 30 minutes. Dissolve the residue in 10 ml of 10 N sulphuric acid, and warm for 15-20 minutes. Add 20 ml of water and filter through Whatman No. 1 filter paper into a 100 ml volumetric flask. Wash the filter paper several times with hot water and dilute to volume.

DEVELOPMENT OF COLOUR

Pipette an aliquot (5-15 ml) of ash solution into a 50 ml volumetric flask. Add a calculated* amount of 10 N sulphuric acid to make the amount of acid equivalent to 2.0 ml of 10 N sulphuric acid. Dilute to about 45 ml, add 1 ml of ammonium molybdate solution and shake well. Add one ml of stannous chloride solution and shake immediately. Dilute to volume and shake. After 20 minutes read the percentage transmission in a colorimeter at 660 mμ.

CALIBRATION CURVE

Pipette several aliquots (0.5 ml) of working standard phosphate solution into 50 ml volumetric flasks. Proceed exactly as in case of test solution. 1 ml of working standard represents 0.01 mg of phosphorus.

$$\%\text{ of Phosphorus} = \frac{\text{amount of phosphorus}}{\text{in the aliquot, g}} \times \frac{\text{aliquot}}{\text{factor}} \times \frac{100}{\text{wt. of sample, g}}$$

*The 50 ml blue coloured solution should be 0.4 N with respect to sulphuric acid, to allow maximum colour development.

1.72 Amount Present in Soil and Water

Cations are removed from the exchange complex by leaching the soil with ammonium acetate solution. Various cations are estimated separately in aliquots of the soil extract.

Procedure A (Exchangeable base):

REAGENTS

(*i*) 40% ethyl alcohol. Add 600 ml of water to 400 ml of absolute or 95% alcohol.

(*ii*) Absolute alcohol.

(*iii*) N ammonium acetate solution. Prepare separately 2N solutions of acetic acid and ammonia. For 2N acetic acid dilute 115 ml of glacial acetic acid (sp. gr. 1.052) to 1 litre. In case of ammonia dilute 108 ml (sp. gr. 0.88) or 150 ml (sp. gr. 0.91) of ammonium hydroxide to 1 litre. Mix equal volumes of both the solutions. Adjust the pH of the ammonium acetate solution to exactly 7 by adding either 2N acetic acid or ammonia.

Transfer 50 g of air dry soil sample into a 500 ml beaker. Add 100 ml of 40% alcohol, stir well and allow to stand for 15 minutes. Pour the suspension into a Buchner funnel fitted with Whatman No. 50 filter paper. Wash the soil in the same way with 50 ml portions of 40% alcohol 4-5 times and finally with 50 ml of absolute alcohol to dry the soil. Washing with alcohol may be omitted if the soil contains small amounts of soluble salts. Return the soil to the beaker, washing the funnel and the filter paper with 100 ml of a ammonium acetate solution. Stir the soil suspension for a few minutes and allow to stand overnight. Decant the supernatant into a Buchner funnel fitted with Whatman No. 42 filter paper and collect the filtrate into a 500 ml volumetric flask. Transfer the soil into the funnel with the help of a jet of ammonium acetate solution. Continue leaching with successive additions of 50-60 ml of ammonium acetate solution, allowing the soil to drain completely after each addition.

Procedure B (Calcium): Calcium is precipitated as calcium oxalate and the latter is dissolved in sulphuric acid and titrated against potassium permanganate.

REAGENTS

(*i*) 6N Hydrochloric acid. Dilute 250 ml of concentrated acid to 500 ml.

(*ii*) Glacial acetic acid.

(*iii*) Brom-cresol green indicator. Prepare 0.5% alcoholic solution.

(*iv*) Ammonium hydroxide, concentrated.

(*v*) 4N ammonium hydroxide. Dilute 130 ml concentrated ammonium hydroxide to 500 ml.

(*vi*) 10% Oxalic acid. Dissolve 50 g in water and dilute to 500 ml.

(*vii*) 2N Sulphuric acid. Carefully pour 55 ml of concentrated acid in 800 ml of water and dilute to 1 litre.

(*viii*) 0.05N potassium permanganate. Bring approximately 1 litre of water to boil and add 1.6 g of potassium permanganate. Continue boiling for 15-20 minutes and restore the original volume. When quite cold filter (preferably through sintered glass crucible) and store in amber coloured bottle. Standardise against oxalic acid before use.

Prepare standard solution of oxalic acid (0.05 N) by dissolving exactly 3.151 g pure dry acid in water, adding 20 ml of concentrated sulphuric acid and when cold diluting to 1 litre. Pipette 20 ml of standard oxalic acid solution in a beaker, add 5 ml of concentrated sulphuric acid and heat to 80-85°C. Titrate with potassium permanganate, with rapid stirring, till pink colour just persists.

$$\text{Normality of permanganate} = \frac{\text{Volume of oxalic acid} \times \text{normality of oxalic acid}}{\text{volume of permanganate used}}$$

Place an aliquot (100-200 ml) of ammonium acetate extract in 200 ml basin and evaporate to dryness on a waterbath. Then ignite at dull red heat for 10 minutes to decompose ammonium acetate. Cool and add 10 ml of 6 N hydrochloric acid followed by 15 ml of water. Stir to dissolve completely and filter through a Whatman No. 42 filter paper. Wash with hot water and collect the filtrate in a 250 ml beaker. Add successively 5 ml 6N hydrochloric acid, 5 ml glacial acetic acid and 0.5 ml bromcresol green indicator to the filtrate. Neutralise with concentrated ammonium hydroxide to greenish tinge; bring to boil; and add 10 ml of 10% oxalic acid. Add 4N ammonium hydroxide drop by drop, with constant stirring, till the solution becomes distinctly green. Heat on a water bath for 1 hour and pour the supernatant into a fritted glass crucible or funnel fitted with Whatman No. 42 filter paper. Transfer the precipitate to the filter and thoroughly rinse the beaker thrice with hot water. Pour the rinsings through the filter and wash the precipitate at least 5 times with small amounts of hot water. Collect the filtrate in a 500 ml beaker and save it for magnesium estimation.

Place the fritted glass crucible with precipitate in the original beaker and add 75 ml of 2 N sulphuric acid and sufficient water to cover the crucible. Heat the solution to 80-85°C and titrate with potassium permanganate till the pink colour persists. When a filter paper has been used, pierce it and wash the precipitate into the original beaker. Wash the filter paper with 75 ml of 2 N hot sulphuric acid and many times with hot water. Heat the solution to 80-85°C and titrate with permanganate, adding the filter paper just before the end point.

CALCULATIONS

Calculate the results by the formula given below. Meq. means equivalent weight represented in milligrams.

$$\text{Meq. of exchangeable Ca/100 g soil} = (\text{ml permanganate}) \times (\text{permanganate normality}) \times (\text{aliquot factor}) \times \frac{100}{\text{wt of soil g}}$$

If 100 ml aliquot is used the factor will be $\frac{500}{100} = 5$

If the soil contains more than 0.3% calcium carbonate deduct the amount of calcium present as carbonate from the value of exchangeable calcium. Calcareous soils may be leached with barium chloride-triethanolamine, as originally outlined by Mehlich (1945) and also discribed by Jackson (p.88, 1962).

REFERENCE

Jackson, M. L. 1962. *Soil Chemical Analysis.* Asia Publishing House, Bombay.

Mehlich, A. 1945. *Soil Sci.* **60:** 289.

Procedure C (Magnesium): Calcium free solution is reacted with sodium phosphate to precipitate magnesium. The precipitate ($Mg\,NH_4PO_4$) is ignited and weighed as pyrophosphate ($Mg_2P_2O_7$).

REAGENTS

(*i*) 10% Ammonium phosphate [$(NH_4)_2HPO_4$]. Dissolve 10 g in water and dilute to 100 ml.

(*ii*) Ammonium hydroxide (concentrated).

(*iii*) Dilute ammonium hydroxide. Dilute concentrated solution 10 times.

(*iv*) 0.04% Bromocresol purple indicator. Powder 0.4 g of indicator with 7.4 ml of 0.1 N sodium hydroxide solution in a glass mortar. After an hour transfer the solution in a 2 litre flask, dilute with water to 1 litre and adjust the colour to purple by dropwise addition of dilute hydrochloric acid.

To the filtrate from calcium estimation add a few drops of bromo-cresol purple indicator and add ammonium hydroxide till the solution becomes purple. Heat the solution on a waterbath to reduce the volume to 100-150 ml. Cool and gradually add 10 ml of 10% of ammonium phosphate with constant stirring. Stir vigorously for a few minutes. Add 30 ml of concentrated ammonium hydroxide with rapid stirring. Allow to stand overnight. Filter through Whatman No. 42 filter paper. Rinse the beaker 3-4 times with dilute ammonium hydroxide, rubbing the sides with a rubber-headed glass rod, to transfer the precipitate quantitatively to the filter paper. Wash the precipitate 5-6 times with dilute ammonium hydroxide. Ignite the filter paper containing precipitate in a weighed silica crucible till a white residue is obtained. Cool in a desiccator and weigh.

CALCULATION

Multiply the weight of precipitate in mg by 0.218 to find the weight of magnesium.

Meq. exchangeable Mg/100 g soil =

$$\left(\frac{\text{wt. of Mg in the ppt., mg}}{24.3}\right) \times (\text{aliquot factor}) \times \left(\frac{100}{\text{wt. of soil, g}}\right)$$

Procedure D (Potassium): Potassium is precipitated as cobalt nitrate complex $(K_n\ Na_{3-n})\ Co\ (NO_2)_6 . xH_2O$. The precipitate is dissolved in 2 N sulphuric acid in presence of an excess of potassium permanganate and the remaining permanganate is back titrated with oxalic acid. Concentration of potassium in the unknown is determined with reference to a standard curve.

REAGENTS

(*i*) 10% Sodium hydroxide.

(*ii*) 1 N hydrochloric acid (ammonium free). Dilute concentrated acid ten times.

(*iii*) Phenolphthalein indicator (1% alcoholic solution).

(*iv*) 20% sodium cobaltinitrite solution. Dissolve 20g of cobaltinitrite $[Na_3\ Co\ (NO_2)_6]$ in 70-80 ml of water previously cooled to 5°C. Allow to stand overnight and filter through Whatman No. 42 filter paper. Store the reagent in tightly stoppered bottle in a refrigerator.

(*v*) 0.17N acetic acid containing 4% formaldehyde. Mix 10 ml of glacial acetic acid and 100 ml of 40% formaldehyde and dilute to 1 litre.

(*vi*) Standard potassium chloride solution. Dissolve exactly 0.1905 g of pure potassium chloride in water and dilute to 500 ml. 1 ml contains 0.2 mg of potassium.

(*vii*) 70% of absolute alcohol.

(*viii*) 0.05 N potassium permanganate (see calcium estimation).

(*ix*) 0.05 N oxalic acid (see calcium estimation).

(*x*) Concentrated, sulphuric acid.

Evaporate to dryness and ignite a suitable aliquot (100-150 ml) of ammonium acetate extract, as in case of calcium. Cool and wash down the sides of the basin with a jet of water. Add a drop of phenolphthalein indicator and 2-4 drops of 10% sodium hydroxide. Evaporate to dryness to remove last traces of ammonia. Take up the residue in 2 ml of 1 N hydrochloric acid and again evaporate to dryness. Cool and dissolve the residue in exactly 15 ml of 0.17 N acetic acid, rubbing the basin with rubber-headed glass rod. Immediately filter through a dry filter paper into a 50 ml conical flask and tightly stopper the latter. Cool the flask to 3°C.

In the meanwhile pipette 10 ml of 20% sodium cobaltinitrite solution into a 50 ml beaker and cool it to 3°C. When both solutions have cooled to 3°C, pipette 10 ml of potassium solution into the beaker containing sodium cobaltinitrite solution, with constant rapid stirring by rotating the beaker. Allow to stand at 4° to 6°C for 5 hours, preferably overnight.

Filter through a Gooch crucible charged with asbestos,* washing the beaker thrice with 70% alcohol. Dry in an oven at 100—110°C for about 20 minutes.

Pipette 25 ml of 0.05 N potassium permanganate in a 250 ml beaker, dilute to about 150 ml and add 5 ml of concentrated sulphuric acid. Then add to the crucible with precipitate, stir and warm gently. If the solution begins to turn colourless, immediately add measured amount of potassium permanganate to ensure an excess. Heat to boiling, remove and rinse the crucible. Add an excess of 0.05 N oxalic acid and warm till the solution becomes colourless. Titrate against 0.05 N potassium permanganate.

---

*Before use asbestos should be digested in aqua regia (HCl 3part + HNO_3 1 part) at boiling temperature for 10 minutes and then washed well with water.

STANDARDIZATION

Evaporate 15 ml aliquot of standard potassium chloride solution and proceed exactly as described above. This aliquot represents 2 mg potassium, because out of 15 ml potassium solution only 10 ml is used for precipitation.

CALCULATIONS

Amount of permanganate used up to oxidise the precipitate is calculated as given below.

ml of permanganate = Total ml of permanganate − ml of oxalic acid utilized × R

$$\text{Where } R = \text{ratio of } \frac{\text{(Permanganate volume)}}{\text{(oxalic acid volume)}}$$

Generally strengths of permanganate and oxalic acid are not exactly equal. There the strength is determined by titrating 10 ml of hot oxalic acid against permanganate.

The amount of potassium is calculated as follows:

$$\text{mg potassium in test solution} = \frac{\text{mg potassium in test solution}}{39.1} \times \text{aliquot factor} \times \frac{100}{\text{wt. of soil, g}}.$$

PRECAUTION

The laboratory must be completely free from fumes of ammonia during estimation.

The methods of estimation of calcium, magnesium and potassium outlined above are based upon those described by Jackson (1962). Many alternative procedures are also described by the same author. In recent years flame photometric estimation of potassium has come into general use. A step further, with the help of a spectrophotometer, having flame attachment, calcium, magnesium and also potassium can be estimated. These instruments permit accurate estimation of large number of samples in a short time (if a few samples are to be analysed chemical methods are more convenient). Mehlich (1945) has described a convenient scheme for volumetric and flame photometric estimation of cations and anions.

REFERENCES

Jackson, M. L. 1962. *Soil Chemical Analysis*, Asia Publishing House, Bombay.

Mehlich, A. 1945. *Soil Sci.* **60:** 289.

Procedure E (Available phosphorus): Dilute acids remove a fraction of total phosphorus from the soil which has been called "available" phosphorus. Phosphorus reacts with molybdate to form yellow coloured heteropoly-complexes. Reduction of heteropoly-complexes to give blue colour, forms the basis of colorimetric estimation of phosphorus.

REAGENTS

(i) 0.002 N sulphuric acid. Dilute concentrated acid to prepare approximately 0.1 N sulphuric acid. Standardise against standard (0.1 N) sodium carbonate solution. Dilute suitable amount to give exactly 0.002 N solution. Before making up the volume add ammonium sulphate or potassium sulphate at the rate of 3 g per litre.

(ii) Sulfomolybdic acid solution. Dissolve 25.0 g of ammonium molybdate $[(NH_4)_6 Mo_7O_2 . 4H_2O]$ in 200 ml of water and heat to 60°C. Separately dilute 275 ml of concentrated sulphuric acid to 750 ml. Cool both solutions and then mix them. When cool dilute to 1 litre. Store in dark coloured bottle.

(iii) 0.1 M Chlorostannous acid. Dissolve 25 g stannous chloride ($SnCl_2 2H_2O$) in 50 ml. of concentrated hydrochloric acid. Dilute to approximately 500 ml with water and finally make up the volume to 1 litre with 1.2 N hydrochloric acid. Chlorostannous acid solution should be prepared immediately before use.

(iv) Standard phosphate solution. Dissolve 0.2195 g pure potassium dihydrogen phosphate (KH_2PO_4) in water. Add 25 ml of 7N sulphuric acid and dilute to 1 litre. This gives 50 ppm stock solution of phosphorus. To prepare working standards dilute 2, 5, 10, 15 and 20 ppm solutions respectively.

(v) 0.25% 2, 4 dinitrophenol indicator. Dissolve 0.25 g in 100 ml water.

EXTRACTION OF PHOSPHORUS

Transfer 1.00 g of soil to a 500 ml conical flask and add 200 ml of 0.002 N sulphuric acid. Mechanically shake the suspension for 30 minutes and then filter through a Whatman No. 42 filter paper.

DEVELOPMEMT OF COLOUR

Pipette an aliquot (10-25 ml) in a 50 ml volumetric flask and add 2-3 drops of dinitrophenol indicator. If the solution becomes yellow, add 2N sulphuric acid drop by drop till it turns colourless (pH_3). If the solution remains colourless after addition of indicator, add 4N sodium carbonate solution drop by drop till it becomes yellow. Then add 2N sulphuric acid till it becomes colourless.

Dilute the solution to about 4.5-4.7 ml and add 2 ml of sulphomolybdic acid solution. Shake and add 0.5 ml of chlorostannous acid. Shake and make up to volume. Shake well and read the% transmission at 660 mμ after 5 minutes.

CALIBRATION CURVE

Pipette an aliquot (same as the test solution) of 2 to 20 ppm standard solutions in 50 ml volumetric flasks. Develop colour exactly as in case of test solutions. Plot transmission percentages against concentration of standards.

CALCULATIONS

Find out the concentration of test solution by referring to the calibration curve.

$$\text{ppm phosphorus in soil} = \frac{\text{ppm phosphorus in solution}}{\text{(from calibration curve)}} \times \frac{\text{ml. extraction solution used}}{\text{wt. of soil g}}$$

When 1 g of soil has been extracted with 200 ml solution, ppm phosphorus in soil=ppm phosphorus in solution $\times$ 200.

REFERENCE

Jackson M. L. 1962. *Soil Chemical Analysis.* Asia Publishing House, Bombay.

NOTE:— From the data of the foregoing exercises a mineral budget can be prepared, showing clearly the amount of them released and conserved during a year per unit area of grassland.

1.80 SUCCESSION

With the development of plant community through migration of propagules from adjoining areas, the substratum becomes variously modified as a result of plant-soil reaction and death and decay of the species growing thereon. This modification makes the habitat unsuitable for the existing plant community and offers a congenial habitat to fresh invaders. Thus a new plant community develops on the same site. This process may result in the development of a number of plant communities, one after another at the same site. This phenomenon is termed as succession and involves the processes of migration, ecesis, competition, reaction and stabilization. It may be noted here that the invasions may not be obvious for sometime because of the dominance of already existing plant species. But the mutual effects of plants and habitat upon each other soon cause elimination of some of the already existing individuals lessening the competition, and the invaders start flourishing. Some of the invaders have strong competitive power and may oust the species through the process of competition alone.

The succession may be primary or secondary depending upon whether it starts on a bare area where plants never existed before or on disturbed (such as burnt, flooded, or scraped) areas respectively. In both cases it leads to diversity of species even though the biomass may remain the same or decline sometimes.

1.81 Recording the Data From the Sites Showing Different Stages of Succession

Procedure A (Secondary succession): Select the areas in a grassland representing different successional status; say the areas where grasses and forbs are just making appearance, where they have developed further, and finally the area which corresponds to undisturbed piece of grassland in that particular season. It is desirable to select anumber of sites intermediate between disturbed and undisturbed ones. Make separate lists of the species growing on each site. Also estimate the frequency, density and cover of the plant species in each case following the quadrat method described in exercises 1.11/A, 1.12/A, and 1.13/A respectively.

Record your data in the following table:

Plant species	Seral stages											
	A			B			C			D		
	F	D	Co	F	D	Co	F	D	Co	F	D	Co
Species a												
Species b												

F=Frequency, D=Density, Co=Cover

INTERPRETATION

On the basis of the data set in the above table indicate the plant species predominating on different sites.

Compare the values for frequency, density and cover of the plant species common to all the sites indicating the trend of increase or decrease in these values.

Note whether some of the species (also name them) present in site 'A' become important components of an undisturbed piece of grassland. List the species present on the disturbed site which are sufficiently suppressed or completely eliminated from the undisturbed area. Indicate the successional trend in the grassland under study.

REFERENCES

Curtis, J. T. 1959. *Plant Ecology Work Book: Laboratory, Field and Reference Manual.* Burgess Publishing Co., Minnesota.

Oosting, H. J. 1958. *The Study of Plant Communities.* W. H. Freeman & Co., San Fransisco.

Weaver, J. E. and F. E. Clements. 1938. *Plant Ecology.* McGraw Hill Book Co., New York.

Procedure B : Maintain exclosures by enclosing small areas in a grassland by means of wire-netting. The enclosures of different ages may be obtained by periodical setting. Examine the botanical composition of different enclosures. Find out the density frequency and cover of the species present in each and tabulate your data as indicated below:

Plant species	Exclosures (young to old)											
	A			B			C			D		
	F	D	Co	F	D	Co	F	D	Co	F	D	Co
Species 'a'												
Species 'b'												
Species 'c'												

F=frequency, D=Density, Co=cover

INTERPRETATIONS

Interpret your results as in the previous exercise. Also note and compare the number of annual and perennial plant species in young and old enclosures.

REFERENCE

Phillips, E. A. 1959. *Methods of Vegetation Study.* Henry Holt & Co., Inc.

Procedure C: Vegetation is dynamic and is characterised by constant changes in its structure. There are a number of causal factors (including season). Every season with a definite set of climate conditions supports the growth of certain definite plant species, and thus, with the change in season several vegetational changes (termed as seasonal) are brought about.

Lay the permanent quadrats in a grassland. Find out the cover of each species by following the method outlined in exercise 1.13/A. Also estimate the density of the plant species encountered. Protect the quadrat from biotic disturbances and take bimonthly observations in the same quadrat. Record your data in the manner indicated below:

Plant species	Rainy season		Winter season		Summer season	
	July D C	Sept. D C	Nov. D C	Jan. D C	March D C	May D C

D=Density, C=Cover

INTERPRETATION

(1) Compare the botanical composition of the grassland in different seasons.

(2) Plot the seasonal changes in the values of density and cover of the plant species on a graph paper.

(3) Name the dominant species of the three seasons.

(4) Also note the phenological behaviour of the species.

(5) How far such seasonal changes can be interpreted as succession? Note that succession is developmental and seasonal changes are recurrent.

Procedure D: Unlike the previous exercise here it is unnecessary to maintain a permanent quadrat. Seasonal survey of the grassland and the estimation of density and cover of the species as in previous experiment will serve the purpose. It may be mentioned that the grassland under study is open to all sorts of biotic disturbances with no protection. Tabulate the data as in previous exercise.

INTERPRETATION

(1) Interpret the results obtained as above.

(2) Compare the results of this experiment with those of the previous experiment. Assign reason for the differences, if any.

(3) Repeat this experiment next year and see whether seasonal changes are repeated with negligible differences.

(4) Compare and contrast the seasonal changes with those brought about through succession.

1.82 Energy Level of Standing Crop of the Sere

Exercise 1.81 gives a method to ascertain the seral stages. Ascertain various seral stages in a grassland and determine the total energy of the standing crop per unit area by the method already described in Exercise 1.32.

Record your data as follows:

S.No.	Seral stage	Energy per unit area

INTERPRETATIONS

With the progress of succession the food chain relationships become more and more complex. The percentage loss of energy due to respiration is greater for higher trophic levels. For producers it is nearly 33% whereas for primary consumers it is 62% and becomes still higher for secondary consumers. Similarly it has been seen that the gross production increases as the succession proceeds but for the ecosystem as a whole P/R ratio changes from P/R $\gtrless$ 1 to P=R, when the climax is attained.

1.83 Succession Diagram

The succession can be visualised by any of the three following Community characteristics:

(1) Species structure,

(2) Organic structure, and

(3) Energy flow (community metabolism).

Taking into account any one of these characteristics the succession diagram can be drawn. So far the importance has been attached to the species structure of the community and the successional diagrams have been drawn on that basis.

Procedure A: The data accumulated in exercise 1.81 would give a clear idea of the seral stages; these stages are put in order from the pioneer to the climax and their relations are shown by arrows. The pioneer community is placed at the bottom and the climax community is placed at the top.

Procedure B: Compute the total energy of the standing crop at the various seral stages (c.f. exercise 1.82). Plot the productivity against time.

OBSERVATIONS

A sigmoid curve, with small undulations at various levels is obtained.

INTERPRETATIONS

Hutchinson and Wollack (1940) have shown that changes during succession in the rate of production may be expressed as a sigmoid curve, showing a rough resemblance to the growth curve of an organism.

Lindeman (1942) has indicated that this curve will contain undulations corresponding to seres. This undulation represents the persistent influence of some combinations of limiting factors, which, until they are overcome by species substitution etc., tend to decrease the acceleration of productivity and maintain it at a more constant rate.

At the climax stage it assumes a constant rate.

REFERENCES

Hutchinson, G. E. and Anne Wollack. 1940. Studies on Connecticut Lake Sediments. II. Chemical analysis of a core from Linsley pond, North Branford. *Amer. Jour. Sci.* **238:** 493-517.

Lindeman, R. L. 1942. The trophic-dynamic aspect of ecology. *Ecology* **23:** 399-418.

Misra, R. 1946. A study in the ecology of low-lying lands. *Indian Ecologist* **1:** 1-20.

Misra, R. and N. K. Joshi. 1952. The forest-complex of Patharia hill, Sagar, *J. Indian bot. Soc.* **31:** 154-170.

Odum, E. P. 1959. *Fundamentals of Ecology.* W. B. Saunders Co., U.S.A.

Odum, E. P. 1963. *Ecology.* Holt. Rinehart and Winston, Inc.

2. FOREST ECOSYSTEM

The forest ecosystem consists of a great variety of living organisms which differ significantly in shape, size and number. The principal plants in forest ecosystem are trees which have been the dominant plants over much of the land surface for thousands of years (Ovington 1965). Though forests extend from the Arctic to the Antarctic circles, the ecological conditions which vary so much from place to place, produce a variety of combinations at various stages of development. The existence of natural forests is possible only above 25 I (Pateron-Weck Index) and the growth is most luxuriant at about 30,000 I (Ovington 1965). The Pateron-Weck Index is calculated through the following formula:

$$I=\frac{Tv}{Ta}\times\frac{P}{1}\times\frac{G}{12}\times\frac{E}{100}$$

where, Ta=mean monthly temperature (in°C) of the warmest month, Tv=difference (in °C) between mean monthly temperature of the warmest and coolest months, P=mean annual precipitation (in mm), G=length of growing season (in months) and $E=\frac{100Rp}{Rs}$ where Rp=radiation at the pole and Rs=radiation at the site (both radiation values being 10^3 g cal/cm^2/min—^{1}).

REFERENCE

Ovington, J. D. 1965. *Woodlands.* The English Universities Press Ltd., London.

2.00 HABITAT STUDY

2.01 Mapping of the Site

(Refer 1.01)

2.02 Physical Characters of Soil

((Refer 1.02)

2.03 Chemical Characters of Soil

(Refer 1.03)

2.04 Atmosphere

(Refer 1.04)

2.05 Biotic

(Refer 1.05)

2.10 COMMUNITY STRUCTURE

A forest ecosystem is characterised by the predominance of trees and shrubs. Herbaceous species are relatively less important as far as the apparent structure of a forest is concerned. However, herbs form the ground flora community, the structure of which will have to be considered separately.

For the quantitative analysis of the forest structure all the three methods described in grassland ecosystem, viz., quadrat, transect and point centered quarter, may be applied.

2.11 Frequency

Procedure A (Quadrat method) : Here the shape of the quadrat is generally circular and the size is larger. In thick forests the quadrat of 6 m radius is convenient whereas in open forests the radius of the quadrat may be increased to 11.5 m (Misra and Puri 1954).

For procedure, recording of data, calculation and interpretation refer to exercise 1.11/A.

REFERENCE

Misra, R. and G. S. Puri, 1954 *Indian Manual of Plant Ecology.* Oxford Book Co., Calcutta and New Delhi.

Procedure B (Transect method): The steps to be followed are essentially the same as described in exercise 1.11/B. However, the length and width of the belt transect in this case, usually happen to be greater and differ according to the nature of the study, physiographic features of the area and the type of vegetation to be studied. The transect to be laid in a forest may be from a few cm to several meters wide, sufficiently long, sometimes a few kms.

2.12 Density

Procedure A (Quadrat method) : Employ a large quadrat as indicated in the previous exercise. Follow the steps detailed out in exercise 1.12A for finding out the density and abundance of plant species.

2.13 Cover and Basal Area

With the help of the measuring tape measure the circumference of trees encountered in a unit area, say 10×10 meter, at breast height and record the data in a table as given below:

Plant species	Circumference of the individuals in cm								
	1	2	3	5	6	7	8	9	10

CALCULATIONS

From the above data calculate the radius of the bole of the individual trees by using the following expression.

$$r = \frac{\text{Circumference}}{2\pi}$$

where r = radius of the tree and $\mu = 3.14$
After finding out the value of radius, calculate the basal area of a tree as follows:

$$\text{Basal area} = \pi r^2 \text{ or } \frac{C^2}{4\pi}$$

Thus calculate the basal area of all the trees encountered. Sum up the values of the trees of the same species. Now find out the total basal area by summing up the basal areas of all the species.

Calculate the percentage basal area as follows:

$$\%\ \text{basal area} = \frac{\text{Total basal area}}{\text{Area of study}} \times 100$$

INTERPRETATION

List the plant species in order of decreasing dominance.

2.14 Importance Value Index and Other Characteristics

Procedure A: Follow the steps adopted in exercise 1.14/A.

Procedure B: Follow the steps described in exercise 1.14/B. Here the herbaceous species are not taken into consideration, and the distance and diameter measurements of the nearest occurring tree in each quarter are taken.

Procedure C (Ground flora) The structure of the ground flora community of a forest is studied through the method described for the study of community structure of grassland ecosystem. Size of the quadrat and transect, and other details also remain the same.

2.15 Comparison of Stands, Ordination and Continuum

Procedure A (Comparison of two communities): The degree of similarity or difference between any two stands can be determined and expressed through the use of community coefficient which depends on the quantitative phytosociologic characters of species common to both stands.

Prepare table and treat the data as described under exercise 1.15/A.

In this case usually tree and shrubby species are considered. However, herbaceous species may also be considered if emphasis is being laid on micro-climate.

Procedure B (Ordination): When a large number of stands of any forest ecosystem have been examined, test can be made to see if the stands exhibit some pattern of distribution or performance through the technique of ordination.

Details are as in exercise 1.1 /B.

Procedure C (Continuum): In order to study how one stand merges into the other forming a chain-like sequence depending upon the gradient of any environmental factor the performance of dominant species is noted and the concept of continuum is applied.

Details are as in exercise 1.15/C.

2.16 Inter-and Intra-Specific Relations

I. INTERSPECIFIC ASSOCIATION

Procedure A: As in exercise 1.17/IA.

Procedure B (Plot less technique): Goodall (1965) has proposed plot-less tests for interspecific association using the distance measures. The method is especially applicable to species with readily distinguishable individuals. In one of these tests use is made of counts of mutually nearest neighbours and in the other distances from a random point to a species and that between one species to the other.

I. Select a particular individual (of any species) and compare the distance to surrounding individuals of other species and identify its nearest neighbour. Then find out the nearest neighbour for this second individual from among the surrounding species and if this neighbour happens to be the first selected individual designate the two species as mutually nearest neighbours. Thus work out the pairs of nearest neighbours and calculate how many times different species pair together.

Now partition the observed number of individuals into random pairs with the caution that each pair contains individuals of different species. Now count the number of pairs of each type. Repeat this procedure several times and build up an empirical frequency distribution for each type of pair. Now compare the observed numbers for each pair of species with the above empirical sampling distribution and express the results as "the proportion of random samples in which the observed number of pairs, or one more extreme in the same sense occurred". In other words, estimate the one-tail probability of the observed figure. In cases, where observed number of pairs has the medium frequency of the empirical distribution, estimate the probability at 0.5.

II. Lay out sampling areas far apart in order to avoid the inclusion of the same individual more than once in one set of measurement. Mark a point in this area and select an individual of each species, randomly. Now measure the distances from the random point as well as centre points of various randomly selected individuals to the nearest individual of every other species whether in the sampling area or not. Compare the distribution of distances to the nearest individual of a species from the random points with that of distances from the randomly selected individuals of each other species. "If the mean inter-specific distance is significantly smaller than the mean random-poini distance, the two species are positively associated; if significantly larger, they are negatively associated" (Goodall 1965). The test of significance may be performed through t—test. Arrange the data in the following manner:

From	To	Mean distance(m)	Mean distance from point (m)*	Degrees of freedom
A	B			
C	B			
etc.				

*Back- transformed means, viz. $(\frac{1}{n} \frac{u}{\Sigma 1} = 1 \sqrt{\overline{xi)}^2}$

REFERENCE

Goodall, D. W. 1965. Plot-less tests of inter-specific association. *J. Ecol.* **53:** 197—210.

2.17 Raunkiaer's Life-Forms and Biological Spectrum

Raunkiaer (1934) proposed a life form system for the description of vegetation on physiognomic basis. In this system the position of the highest perennating organ or bud which provides an indication of the manner in which the plant survives the unfavourable season, is noted and the plants are classified accordingly into the following life form classes:

Symbol		Life form	Description
P		Phanerophyte	Perennating bud well above ground. The class is further sub-divided into :
MM	*(i)*	Megaphanerophyte	Perennating bud above 30m high
	(ii)	Mesophanerophyte	between 8m and 30m
M	*(iii)*	Microphanerophyte	between 2m and 8m
N	*(iv)*	Nanophanerophyte	under 2m
Ch		Chamaephyte	Herbaceous or low woody plants whose perennating buds are borne just above ground level—up to 0.3m
H		Hemicryptophyte	Perennating buds close to ground (rather half hidden in soil).
		Cryptophyte	Perennating organs below soil surface or water. The class is further subdivided into :
G	*(i)*	Geophyte	Perennating buds underground.
HH	*(ii)*	Helophyte	Marsh plants with perennating buds in water logged mud.
	(iii)	Hydrophyte	Perennating buds beneath the water.
Th		Therophyte	Survival in unfavourable season through seeds or spores; annuals.
S		Stem succulent	
E		Epiphyte	

Traverse the forest stand under study and note the position of perennating buds in all the species. Count the number of species in each life-form class (outlined above). Calculate further the percentage of species coming in each class from the total number of species encountered and prepare the biological spectrum of the area.

INTERPRETATION

Compare the above biological spectrum with that of Raunkiaer's normal spectrum which is based on an analysis of random sample of four hundred species representing the world flora, and discuss the departures. The normal spectrum is given below:

Normal spectrum	No. of species	The percentage distribution of the species among the life forms									
		S	E	MM	M	N	Ch	H	G	HH	Th
	400	1	3	6	17	20	9	27	3	1	13

REFERENCE

Raunkiaer, C. 1934. *The Life Forms of Plants and Statistical Plant Geography.* Oxford University Press.

2.18 Profile Diagram

Profile diagram is a pictorial representation of the structure and vertical stratification of the vegetation. It can be prepared by using (i) single plot method used by Davis and Richards (1933), or (ii) distance method described by Knight (1953).

Procedure A (Single plot method): Mark out a suitable plot in the forest.* Prepare a rough sketch (side view) of the dominant species showing their shape, size and arrangement in the stand. Then proceed to find out the height, crown cover and length of clear bole of individual trees. Finally using a graph paper draw the shape of trees according to scale. Give the scale at the left hand side of the diagram.

*Davis and Richard (1933) and many others have used plots measuring 7.6×61 m (25×200 ft.) in case of dense forests. On the other hand Loveless and Asprey (1957) found that a plot of 5×30 m (16.5×99 ft.) was adequate for scrub-thicket vegetation.

Procedure B (Distance method): Sample 20-30 points in each stand. At each point place two 1-meter sticks at right angles to each other, so as to divide the area around the point into four equal quarters. Measure the distance from the point to the nearest tree in each quarter. Record the following for each of the tree at the point:

(i) basal area at breast height, (ii) height of the tree, (iii) height of the lower leaves of the canopy, (iv) width of the canopy. Draw a rough sketch of the tree showing branching pattern.

Calculate the density of trees (as in case of point-centered quarter method). Next, determine the size of plot the diagram would represent, so as to calculate the number of trees to be shown in the diagram. Suppose the plot size is 7.6×61 m and the density of trees is 4.3 per ares. The area of the plot will be 4.64 ares and hence the number of trees expected in this plot will be 4.64×4.3=20. Select out any 20 trees from the 80−120 actually measured in the stand and draw them to scale on a graph paper.

NOTE:—The author has suggested to increase the size of plot (to say 11m×92m) if the number of trees to be shown in the diagram is very small. According to Knight this method offers advantage over the single plot method, because instead of being confined to a small area, it presents an assemblage of trees, each objectively drawn from a larger area. However, a disadvantage with this method is that the trees shown adjacent to each other in the diagram do not necessarily occur together in the stand.

REFERENCES

Davis, T. A. W. and P. Richards. 1933. The vegetation of Moraballi Creek, British Gulnea: an ecological study of a limited area of tropical rain forest. *J. Ecol.* **21**: 350-384.

Knight, D.H. 1963. A distance method for constructing forest profile diagrams and obtaining structural data. *Trop. Ecol.* **4**: 89-94.

Loveless, A.R. and G. E. Asprey. 1957. The dry evergreen formations of Jamaica. *J.Ecol.* **54**: 799-822.

2.20 BEHAVIOUR OF COMPONENT SPECIES (AUTECOLOGY) OF THE ECOSYSTEM

(Refer 1.20)

2.21 Geography, Evolution and Systematics

I. GEOGRAPHY AND EVOLUTION

(Refer 1.21/I)

II. SYSTEMATICS

(Refer 1.21/II)

III. DISPERSAL

(Refer 1.21/III)

2.22 Structural and Physiological Characters and Their Variations

I. STRUCTURAL CHARACTERS

(Refer 1.22/I—A, B, C)

Procedure D (Diameter of the tree): The diameter of a standing tree is most commonly measured at breast height (d.b.h.)—1.5 m above the ground. Large increase in diameter may be measured by callipers or tape. For measuring small amount of growth over short period of intervals, dendrographs and dendrometers are used. The dendrographs record the true growth on a chart mounted on a drum attached to a clock. In case of dendromet ers, the worker has to note readings from time to time.

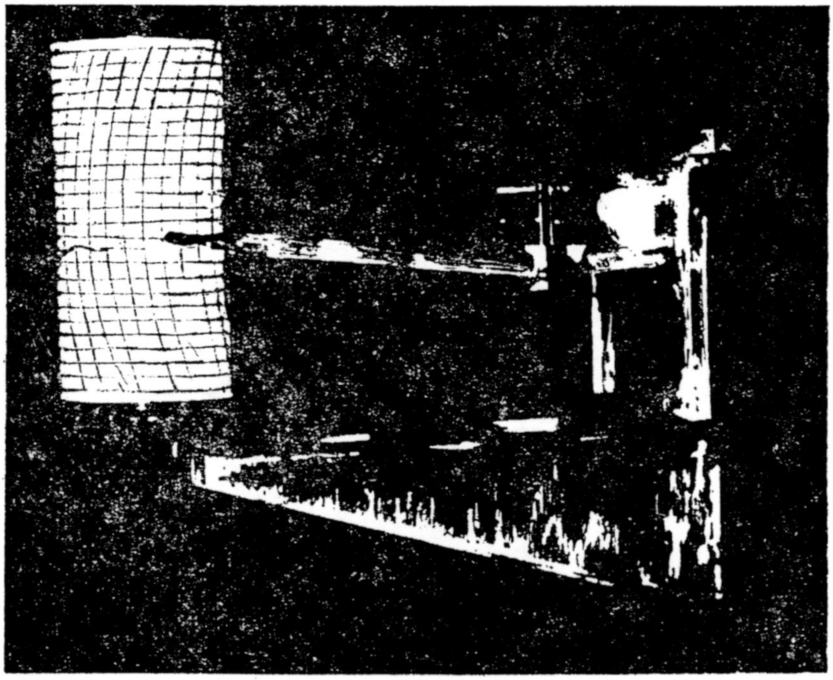

A dendrograph for recording radial growth of tree trunks. (*Fritts and Fritts*, 1955. *Photograph courtesy of H. C. Fritts.*)

FIG. 2.22/ID—1

1) Dendrograph: Fix the dendrograph on the tree trunk as shown in the figure. The chart papers should be replaced periodically and the increase in diameter may be plotted on a graph.

Correlate the growth with moisture and temperature during the respective period.

REFERENCES

Fritts, H.C. and E.C. Fritts. 1955. A new dendrograph for recording radial changes of a tree. *Forest Sci.* **1:** 271-76.

Kramer, P.J. and T.T. Kozlowski. 1960. *Physiology of Trees.* McGraw Hill Book Company, Inc., New York.

2) Dendrometer:

NOTE:—Radial increase in diameter is not even on all sides of a tree. Therefore, several radii at the same height above the ground may be selected and average diameter growth and an estimate of its error may be calculated.

(a) The tips of the 3 wood screws long enough to remain about 2.0 cm above the surface are screwed through the bark and about 1.0 cm of the wood of a tree trunk as shown in figure. The heads of these screws define a plane fixed in position with respect to the outer layers of mature wood. A standard dial gauge (Ames Dial Gauge No. 66 sold by B. C. Ames Co., Waltham, Mass.) is fastened permanently to a small

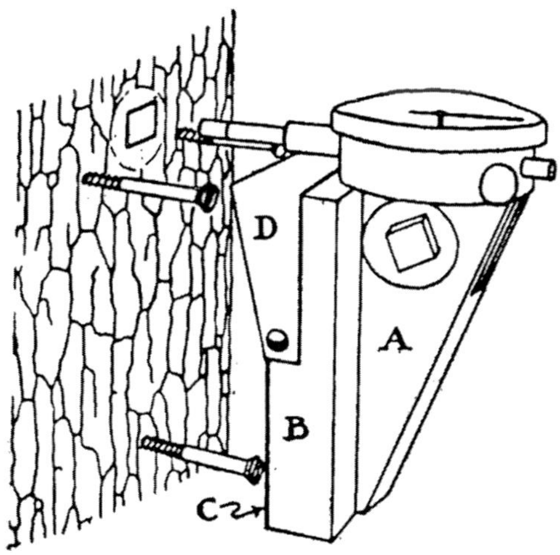

FIG. 2.22/ID-2a

Dendrometer in position just prior to fitting against the 3 screw heads and galvanized metal plate on the bark. Dial gauge is bolted to slotted wooden piece *A*, which is screwed to wooden piece *B*. Surface of *B* is faced with thin steel plate *C*, and partly surrounded by fence *D*.

platform which can be held against the screw heads in making measurements. The spindle of the gauge strikes against a time piece of metal fixed to the bark and the dial registers to 0.001 cm—the distance between the fixed plane defined by the screw heads and the surface of the bark, a distance which decreases as growth adds new tissue between the bark and wood. Only one dial gauge is needed, but each tree under study needs a permanent set of three screws (Daubenmire 1945).

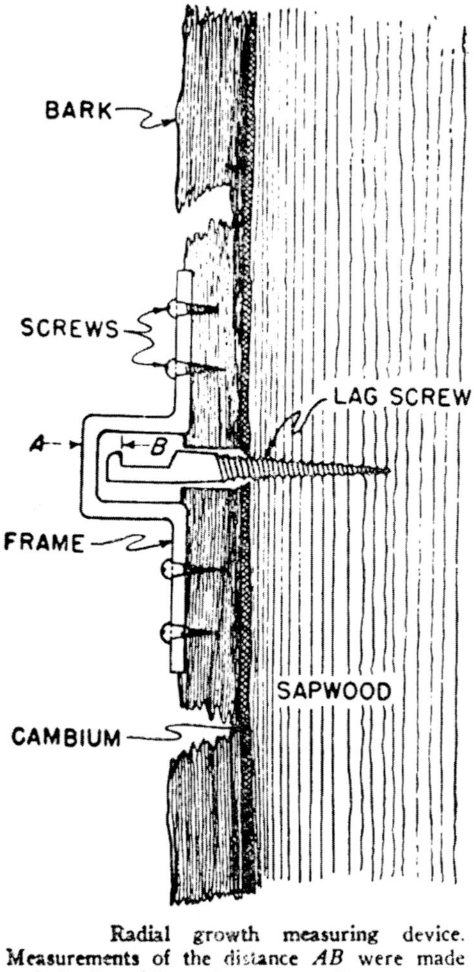

FIG. 2.22/ID-2b

Radial growth measuring device. Measurements of the distance AB were made periodically with a micrometer calipers.

REFERENCE

Daubenmire, R. F. 1945. An improved type of precision dendrometer. *Ecology* **26**: 97-98.

(b) The device consists of an external brass frame screwed tightly to the bark of the tree and a notched steel lag screw inserted through a hole in the bark. The lag screw is screwed well into the sapwood of the tree. Effective lengths of the brass and steel parts are adjusted so that the device is self-compensating for temperature changes. Measurements of the length AB are made with micrometer callipers. Changes in this length as small as 10^{-3} millimeter can be detected. [Byram (1950) found that daily changes were so large that measurements to 10^{-2} millimeter were sufficiently accurate]. Changes in the length of AB are interpreted as changes in the thickness of the cambium layer and neighbouring tissues. Part of this change reflects actual growth and part is a result of the divergence (both positive and negative) of water in living tissues.

Tree ring band for measuring changes in circumference of tree trunks. (*Liming, 1957. U.S. Forest Service photograph.*)

FIG. 2.22/ID-3

REFERENCE

Byram, G. M. and W. T. Doolittle. 1950. A year of growth for a short-leaf pine. *Ecology* **31:** 27-35.

3) Tree-ring band dendrometer : It consists of an aluminium band which encircles a tree. It is held firmly in place by a coil spring. The band is graduated and is fitted with a vernier.

Increase in circumference is noted down at regular intervals.

NOTE:—Try to find out if there is any relationship between available moisture and diameter increase.

REFERENCE

Liming, F. G. 1957. Home-made dendrometers. *Jour. Forestry* **15:** 575-77.

MacDougal, D. T. 1938. *Tree Growth.* Chronica Botanica Co., Waltham, Mass.

Procedure E (Height of the tree) : Two measurements are required with Abney level and forest service hypsometer except when the base or top of the tree is at precisely eye height. They are angular measurements, either from the horizontal as determined by the level bulb on transit and Abney, or from the vertical as determined by plumb bob (or free-swinging weight) in the hypsometer. They must be added or subtracted as shown in the figure.

NOTE:—Abney cannot be accurately used unless the observer is able to stand at least as far away from the tree being measured as its height. Hence it cannot be used in dense stands of very tall timber.

Use of the Abney hand level, if graduated in degrees, parallel that of a transect. If graduated in percentages (the number of cm rise or drop in 30 m of horizontal distance), which correspond to natural tangents, the only computation it requires is adding or subtracting the two readings, pointing off two places and multiplying by the horizontal distance from the tree. If the observations are made from exactly 30 m (horizontally) from the tree, all multiplication can be avoided. If the Abney has a "topographic arc", direct readings are possible only at a distance of 20 m; at 10 m the readings must be halved, at 40 m doubled, etc.

By Staff Hypsometer. It is simply a straight stick 1-1.5 m long. Hold it lightly at arm's length so that the heavy end swings it to a vertical and the length of the stick above the hand equals the distance from hand to eye. Move forward or back until the top of the tree is in line with the top of the stick and the base is in line with the hand.

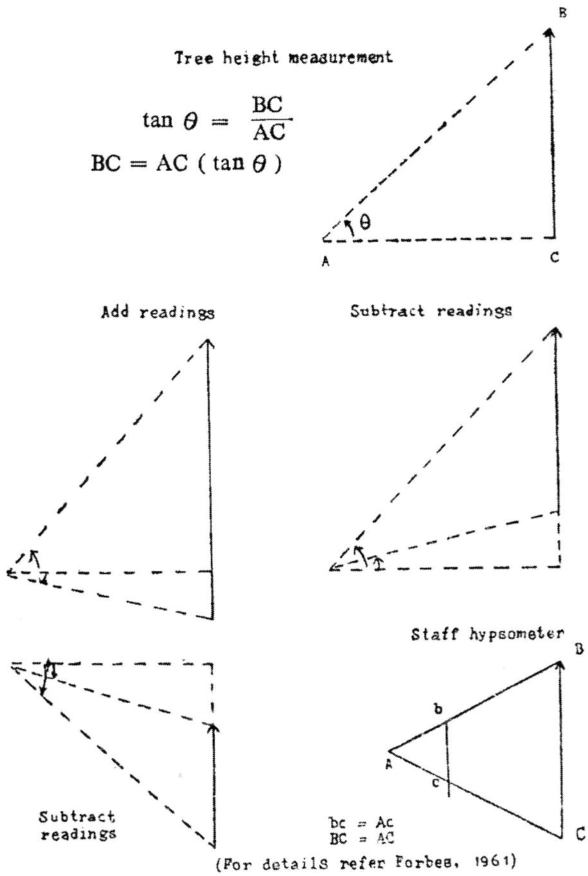

FIG. 2.22/IE

The distance to the base of the tree then equals the tree height.

REFERENCE

Forbes, R. D. (Ed.) 1961. *Forestry Handbook.* The Ronald Press Co., New York.

Procedure F (Flowering zone of a tree): Divide the tree cover in different zones and find out the zone in which the flowers are.

Find out if there is any relation with the insects and birds of this zone with pollination mechanisms.

II. PHYSIOLOGICAL CHARACTERS

(Refer 1.22/II-A*, B, C, D**, E)

Procedure G (Age of the tree): In temperate regions where growth and rest seasons contrast, annual rings in tree stems are well marked. But in tropical regions very few trees such as *Tectona grandis* show clear rings. But even in this tree several rings may arise in a single year. Hence the methods described here as applicable to temperate trees are not trustworthy.

The terms "early wood" and "late wood" are now used in preference to spring-wood and summer-wood (Kramer and Kozlowski 1960).

Take a cross-section of a stem and count the annual rings. Normally one ring is produced in each year by a tree. In normal rings early wood grades into late wood which ends abruptly abutting on the early wood of the following year. False rings may be identified by their poorly defined outer borders.

The young trees form annual rings only after the height growth begins. Therefore, the ring counts do not account for the years the seedlings remained in the sapling stage. (Pessin 1934).

Experience and historical records alone can help in determining the age of a tree in tropical regions. Sometimes correlations between growth performance (e.g. d b h) and age are useful.

REFERENCES

Kramer, P. J. and T. T. Kozlowski. 1960. *Physiology of Trees.* McGraw Hill Book Co., Inc., New York.

Pessin, L. J. 1934. *Amer. J. Bot.* **21:** 599-604.

---

*The flowering and fruiting must be observed round the year.

**Refer the following papers:

Downs, R. J. and H. A. Borthwick. 1956. Effects of photoperiod on growth of trees. *Bot. Gaz.* **117:** 310-26.

Kramer, P. J. 1936. The effect of variation in length of day on growth and dormancy of trees. *Plant Physiol.* **11:** 127-37.

Nitsch, J. P. 1957. Photoperiodism in woody plants. *Proc. Amer. Soc. Hort. Sci.* **70:** 526-44.

Pauley, S. S. 1958. Photoperiodism in relation to tree improvement. In "*The Physiology of Forest Trees*" (Ed. K. V. Thimann). pp. 557-71.

Wareing, P. F. 1956. Photoperiodism in woody plants. *Ann. Rev. Plant Physiol.* **7:** 191-214.

Procedure H (Growth periodicity): Trees manifest diurnal, seasonal and life-time growth periodicities on account of variations in the environment which affect the biochemical processes leading to growth. Baker (1950) has shown that growth of a tree shows a sigmoid curve. Seedlings show a short-period of acceleration. There is very fast growth in young trees. Old trees have a long period of negligible height growth. The rates of shoot growth vary with the species.

Record the shoot elongation of a tree seedling during day and night throughout the year.

Calculate the average day and night growth. See if the rates vary during rainy, winter and summer seasons.

Plot the increase in height on graph sheet.

Find out if there is any relation of height growth with temperature and moisture. (Note the recurrent flushes of growth. Generally the shoots elongate intermittently). Also see if there is any relation between height growth and diameter growth.

REFERENCES

Baker, F. S. 1950. *Principles of Silviculture.* McGraw Hill Book Company, New York.

Nitsch, J. P. 1957. Growth response of woody plants to photoperiodic stimuli. *Am. Soc. Hort. Sci. Proc.* **70:** 512-25.

——1957. Photoperiodism in woody plants. *Ibid.* **70:** 526-44.

2.23 Population Characteristics and Ecotypic Differentiation

(Refer 1.23)

2.24 Cytogenetical and Breeding Behaviour

(Refer 1.24)

2.25 Reproductive Capacity, Dormancy, etc.

I. SEED OUTPUT

In case of trees total output can be calculated as follows:—

(1) Average no. of flowering shoots per tree —X
(2) Average no. of racemes per shoot —Y
(3) Average no. of flowers per raceme —Z

(4) Percentage of flowers producing seeds —A
(5) Average no. of seeds per flower —B

From these data calculate the average seed output of the tree per year.

$$= \frac{X \times Y \times Z \times A \times B}{100}$$

REFERENCES

Sharma, V. B. 1955. An Autecological Study of *Boswellia serrata* Roxb. Ph.D. Thesis, Saugar University, Saugar.

Hatano, K. and S. Asakawa. 1964. Physiological processes in forest tree seeds during maturation storage and germination. *Int. Rev. For. Res.* **1**: 279-323.

II. SEED WEIGHT

(Refer 1.25/II)

III. VIABILITY

(Refer 1.25/III)

IV. GERMINATION

(Refer 1.25/IV)

V. DORMANCY

(Refer 1.25/V)

VI. REPRODUCTIVE CAPACITY

(Refer 1.25/VI)

REFERENCE

Hatano, K. and S. Asakawa. 1964. Physiological processes in forest tree seeds during maturation, storage and germination. *Int. Rev. For. Res.* **1**: 279-323.

2.26 Vegetative Propagation

(Refer 1.26)

In case of forest trees, the regeneration mostly takes place from stumps of felled trees. New shoots are given out from dormant buds at the base of the trunk. But there are a number of such species as *Dalbergia sissoo*, *Pongamia glabra*, *Diospyros melanoxylon*, *Holoptelia*, *integrifolia*, etc. of which the roots yield shoots from underground buds either on exposure due to erosion or on clearing of the surrounding vegetation.

This type of reproduction may be recorded during field trips.

2.27 Seedling Mortality and Growth Rate (Productivity at Different Stages and Seasonal Periodicity)

(Refer 1.27)

Record the conditions for light, temperature and humidity during seedling establishment and growth of the species in nature at the forest floor where the seedlings are found.

Grow the seedlings under controlled conditions and note the effect of above factors upon their establishment and growth.

2.30 FLOW OF ENERGY

2.31 Radiations (Light, Heat Transfer)

Procedure A (Light): To get an idea of input energy in ecosystem the light intensity must be measured in the forests at various strata by the method described in exercises 1.31/A & B.

Procedure B (Radiant energy): To actually compute the total amount of energy available to the producers, the radiant energy has to be measured in the forests at various strata by the method described in exercise 1.31/C and from that the total amount of the energy received at various strata may be computed.

2.32 Annual Increment Due to Producers

In computing the annual increments due to producers, care has to be taken to assess the losses due to respiration, consumers, litter fall and decomposers. In the forests the annual increment due to trees and ground vegetation has to be computed separately per unit area and then the two must be added up to get the final value.

The increment due to ground vegetation has to be computed as in exercises 1.32/A, B & C.

Procedure D (Productivity of tree):

1. Mark out 5 sample plots within the woodland at random.

2. Measure the height, bole diameter at breast height, canopy depth and spread of the trees in sample plots. Find out the averages of each species.

3. Select 2 trees of each species having average height, diameter and canopy characteristics from each plot and cut them down.

4. Cut down each tree into 5 meter pieces from the apex and separate branches, boles, leaves and fruits and weigh separately on a field balance.

5. Transport suitable samples from the above material to the laboratory; dry them in oven to constant weight at 80°C and find out the average.

6. The root system up to 0.5 cm diameter is also excavated and its dry weight determined.

7. This whole operation is done once before the beginning of the growing season, and then at the end of the growing season. The difference between the two gives the annual increment.

8. The average values thus obtained on being multiplied with the total number of trees of that species per hectare will give the annual increment due to producer (tree) per hectare by that species. Similar data obtained and added up for all the species gives the total production.

OBSERVATIONS

Record your observations in the following Table:

S.No.	Species	Oven dry weight					
		Roots	Bole	Branches	Leaves	Fruits	Total

RESULT

Express your results as Kg dry matter produced per unit area per year.

NOTE:— It shall be seen that the dry weight of the leaves and fruits is very small as compared to the other parts and hence their removal by the insignificant consumers will ultimately give a very small percentage of error in computation of total dry matter production. By taking the representative samples of each species and finding out the total energy content per g with the help of a Bomb calorimeter, the total amount of energy conserved per unit area can be calculated.

REFERENCES

Ovington, J. D. and H.A.I. Madgwick. 1959. The growth and composition of natural stands of Birch. I. Dry-matter production. *Plant and Soil* **10:** 271-283.

Ovington, J. D. and W. H. Pearsall. 1956. Production ecology. II. Estimates of average production of trees. *Oikos* **7:** 202-205.

Procedure E (Forest floor): Forest floor is defined as the 'total organic material (including litter) deposited on the mineral soil' (Kittredge 1948). In temperate regions the forest floor is often stratified into three layers; the uppermost layer of relatively unaltered plant and animal remains (L layer), the middle one consisting of fragment and partly decomposed products (F layer) derived from the upper layer and a bottom layer of amorphous finely divided organic matter (H layer) whose source cannot be identified. In tropical well drained soils the above mentioned layers are not obvious and the litter usually rests directly on the mineral soil.

Mark 0.25 m square plot on the forest floor. Cut vertically along the sides. Measure the depth of floor on all four sides. Calculate the average depth of the forest floor. Remove the organic material from the quadrat and weigh. Take sub sample to the laboratory and find out the moisture content. Represent the weight of floor in terms of metric tons per hectare on oven dry basis. Stretch the measuring tape to full length. Record number of half meter segments with or without floor material. Number of half meter segments with floor material represent the percentage of ground covered by forest floor. Replicate this measurement at least 5 times at other places and calculate the average. If necessary reduce the scale, say 10 m length with 10 cm segments.

NOTE:— If a more detailed account is needed, the above measurements may be made separately for individual layers of the forest floor.

REFERENCE

Kittredge, J. 1948 *Forest Influences.* McGraw Hill Book Co. Inc., New York.

Procedure F (Litter Production): Litter is collected in trays placed on soil surface.

In case of evergreen species sampling is done throughout the year whereas, for deciduous species sampling may be restricted to a few months prior and during the leaf-fall.

Place 5-10 trays, distributed at random, in each stand. At monthly or bimonthly intervals, remove the litter collected in each tray and weigh. Carry a sub-sample to the laboratory to determine the moisture content.

Calculate the amount of litter deposited on soil in each month in terms of g per sq. m (oven dried material). Convert these figures to metric tons per hectare. Add up the figures of different months to calculate the annual litter production.

NOTE:— If desired the total litter may be separated into components e.g., twigs, leaves, flowers and fruits. Similarly separation can be made on species basis.

2.33 Food Chain and Rate of Removal of Produce by Consumers

(Refer 1.33)

Estimation of the consumers of various trophic levels is very difficult in forest ecosystem. The whole of food-chain in the woodland community is very difficult to be assessed. In this case as well, the underground and aboveground consumers must be estimated separately.

Divide the tree into different zones, each about 5 m high.

Note down the species of birds and other animals that are detected feeding on the tree in each zone.

Draw a diagrammatic representation of your observation.

In case of birds, insects and other such organisms the population must be determined from time to time by the method given in exercise 1.33/A & B. Place some of these organisms in a cage and find out their consumption per capita in terms of dry matter and with this value calculate out the total consumption of dry matter by the whole population.

The idea of the population of big animals can be gathered by their droppings.

To get an idea of the food these organisms are utilizing, the caught animals may be dissected and their alimentary canal examined.

REFERENCE

Hanson, E. D. 1961. *Animal Diversity.* Prentice-Hall, Inc., New Jersey.

2.34 Energy Released by Decomposers

(Refer 1.34)

2.40 WATER CYCLE

(Refer 1.40)

2.41 Water Input (Precipitation)

(Refer 1.41)

2.42 Return of Water to Atmosphere

I. EVAPORATION

(Refer 1.42/I)

II. TRANSPIRATION

(Refer 1.42/II)

III. CANOPY INTERCEPTION

Interception of rainfall by canopies results in the partition of the incoming precipitation into three components.

C_1 Part retained on the foliage.

C_2 Part channelled down the branches and stems.

C_3 Part falling through unimpeded, or drips from branches.

In trees and woody shrubs simple collection devices as shown in figure, may be attached to the lower part of the stem. The amount of rainfall beneath and outside the canopy is measured at a number of points.

The difference in amount of precipitation recorded outside (P_0) and beneath (C_3) the canopy may be written as:—

$$P_0 - C_3 = C_1 + C_2$$

FIG. 2.42/III: Measurement of canopy interception

For estimating C_2 water collected as shown in the figure should be measured in terms of mm. This can be done by using the following formula.

$$C_2 = \text{Water collected in mm} = \frac{\text{Area of the canopy (as known from the size of shadow at noon)}}{\text{Area of the reservoir of the rain gauge.}}$$

Accordingly $C_1 = P_0 - C_3 - C_2$

The structure of the vegetation is adapted to interception and following the rainfall, the pattern of soil-water recharge is significantly modified. Because of the altered intensity and distribution of the intercepted water and also because of the improved soil permeability caused by the presence of the vegetation the run-off pattern is modified. The portion of the precipitation which is channelled down the trunks of trees, concentrates water, raises the chances of deep infiltration due to high root density and soil permeability at this zone. This leads to the formation of a very favourable situation for the growth of these plants.

2.43 Loss of Water Due to Percolation

(Refer 1.43)

2.44 Loss of Water Due to Run-Off

(Refer 1.44)

2.45 Conservation of Water in the Soil

(Refer 1.45)

2.46 Conservation of Water in the Biomass

(Refer 1.46)

From hydrological point of view any marked change in the normal value of water reserve in different steps of the pathway of water cycle is important. A study of this deviation is of paramount importance to assess the stability of an ecosystem.

The magnitude of the following factors measured periodically throughout the year and for many years, will give us the idea of the amount of water received and conserved by the ecosystem.

(1) Precipitation
(2) Evaporation
(3) Transpiration
(4) Changes in soil water content
(5) Changes in plant water content
(6) Changes in water table.

2.50 CARBON CYCLE

(Refer 1.50)

2.51 Carbon Dioxide Absorption by the Biomass

(Refer 1.32/B)

2.52 Carbon Dioxide Loss From the Biomass

(Refer 1.32/B)

2.53 Carbon Dioxide Loss by Soil and Water

(Refer 1.53)

Forest soils receive considerable amount of organic material through litterfall. Under favourable climatic conditions organic material is decomposed by micro-organisms and is gradually incorporated into the mineral soil beneath. During the process of decomposition CO_2 is abundantly evolved. Amount of organic matter and rate of evolution of CO_2 from the soil can be estimated by the methods described for grassland soils. Decomposition of litter under laboratory conditions can be studied by mixing the litter with soil, bringing the mixture to desired moisture level and following the fate of selected group of compounds by analyzing the mixture periodically (Ivarson and Sowden 1959, Sowden and Ivarson 1959, 1962). Under natural conditions decomposition of tree litter has been studied by Bocock and Gilbert (1957) by employing nylon net technique. This technique originally used for broad leaf species has also been found suitable for needle leaves (Hayes 1965).

Place weighed amount of air dry litter in nylon nets. Calculate the oven dry weight of litter by estimating moisture content in a separate sub sample. Distribute the bags at random on the forest floor and anchor them suitably. Periodically (say after two months) remove a few bags and find out the oven dry weight of litter left in the bag. Estimate the percentage of carbon and nitrogen, in the residual litter on each sampling date, by usual methods.

Follow the loss in dry weight of original litter with time. Similarly find out the changes in carbon and nitrogen content of litter after different degrees of decomposition. Knowing the weight of litter and its carbon and nitrogen contents the changes in absolute amounts of carbon and nitrogen can be calculated.

NOTE:—This method can be used to study the changes in any other constituent of the litter also. For example changes in amount of cold water soluble substances or polyphenols have been studied by different workers.

REFERENCES

Bocock, K. L. and O. J. W. Gilbert. 1957. The disappearance of leaf litter under different woodland conditions. *Plant & Soil.* **9:** 179-185.

Hayes, A. J. 1965. Studies on the decomposition of coniferous leaf litter. I. Physical and chemical changes. *J. Soil. Sci.* **16:** 121-140.

Ivarson, K. C. and F. J. Sowden. 1959. Decomposition of forest litters. 1. Production of ammonia and nitrate nitrogen, changes in microbial population and rate of decomposition. *Plant & Soil* **11:** 237-248.

Sowden, F. J· and K. C. Ivarson. 1959. Decomposition of forest litters. 2. Changes in the nitrogenous constituents. *Plant & Soil* **11:** 249-261.

Sowden, F. J. and K. C. Ivarson. 1962. Decomposition of forest litters. 3. Changes in the carbohydrate constituents. *Plant & Soil* **16:** 359-400.

2.54 Storage of Carbon as Carbonate and Organic Matter in Soil and Water

(Refer 1.54)

CARBON BUDGET

By estimating the annual carbon input of the ecosystem and losses by the different components, the annual carbon budget of the forest can be prepared. The values of carbon are usually expressed as metric tons per hectare. A simplified outline of the budget is given below:

CARBON BUDGET OF FOREST ECOSYSTEM

Components	Gross Carbon fixation	Loss as respiration	Net annual gain or loss
Trees			
Shrubs			
Herbs			
Litter and soil			
Total	A	B	A-B

2.60 NITROGEN CYCLE

Nitrogen circulates within air-soil-autotroph-heterotroph cycle. The original source of nitrogen to the forest soils is microbial nitrogen fixation and via precipitation. Under certain conditions some of the soil nitrogen is lost through denitrification or run-off. On the other hand, most of it is absorbed by vegetation, especially trees. Out of the total amount of nitrogen absorbed annually a small fraction is retained for a long time in tree trunks, while most of it is returned to the soil through litter fall. During litter decomposition most of the nitrogen is conserved in the soil and is made ready for absorption by the vegetation. For estimation of nitrogen in plant material and soils the same general methods as described for the grassland ecosystem are used.

2.61 Nitrogen Input by Nitrogen Fixing Organisms

(Refer 1.61)

NOTE:—In case of leguminous trees which may bear nodules the soil may periodically be collected from the rhizosphere and analysed for its nitrogen contents to estimate the periodical nitrogen fixation by the tree.

2.62 Nitrogen Loss by Denitrifying Organisms

(Refer 1.62)

2.63 Uptake and Storage of Nitrogen in Plants

(Refer 1.63)

NOTE:—In case of trees foliar analysis give a fairly good idea of circulation of minerals.

Nitrogen Budget

Like that of carbon annual nitrogen budget can also be prepared. The basic scheme showing income and losses of nitrogen for the forest ecosystem is presented below:—

INCOME	LOSSES
1. N-content of the soil and standing crop at the beginning of the study.	1. Denitrification.
2. Annual N-fixation from air.	2. Run-off and percolation.
3. Annual N-return from litter fall and root decomposition.	3. N removed by commercial exploitation in boles, twigs and leaves.
4. Total N in annual precipitation.	4. N removed due to grazing.
5. Annual gain from dead bodies of animals.	
Total A	L

Hence, Net annual gain or loss of N by the ecosystem $= A - L$

2.70 CYCLING OF Ca, Mg, P, K

(Refer 1.70)

Nutrients like Ca, Mg, P and K are also circulated between vegetation and soil. Soil originally derives its mineral supply from parent material through the process of weathering. Most of these minerals, released by weathering, are held on exchange complex of the soil, from where they are absorbed by the vegetation. As in case of nitrogen, certain fraction of minerals may be locked up in the trunks for a very long period, while most of them are returned to soil through litter fall. The rate of release of minerals from the litter by decomposition depends upon climate and vegetation type. In warm regions under deciduous forests decomposition is usually completed within an year. But under evergreen forests it may take 4-6 years to release all the minerals.

2.71 Uptake and Storage in Plants

(Refer 1.71)

2.72 Amounts Present in Soil and Water

(Refer 1.72)

NOTE:—From the data of foregoing exercises a mineral budget can be prepared, showing clearly the amount of them released during a year per unit area of forests.

2.80 SUCCESSION

(Refer 1.80)

2.81 Recording the Data From the Sites Showing Different Stages of Succession

(Refer 1.81/A, B, C & D)

Procedure E (Secondary succession through fire) :

Burning destroys the pre-existing vegetation and on burned areas the colonization of plants begins which may again result in the growth of vegetation almost similar to that existing before.

Select a typical area and study its structure through quadrat method and through photography. Burn the selected area and photograph it. Mark the area by means of stakes and rope and record the vegetation after various lapses of time until a condition approximating to pre-burning stage is attained. Record the observations in the following tables:

Table 1

THE STRUCTURE OF THE SELECTED PIECE OF FOREST BEFORE BURNING

Date:

Plant species	Frequency	Density	Cover	Remarks

Table 2

THE STRUCTURE OF THE SAME AREA AFTER BURNING

I. Observation. Date:

Plant species	Frequency	Density	Cover	Remarks

II. Observation. Date:

Plant species	Frequency	Density	Cover	Remarks

III. Observation. Date:

Plant species	Frequency	Density	Cover	Remarks

Final Observation. Date:

Plant species	Frequency	Density	Cover	Remarks

INTERPRETATION

Name the species which initiate secondary succession. What happens with the pioneer species as succession progresses?

How do the new plant species encountered in subsequent observations behave?

Compare and contrast the structure of vegetation before burning with that of finally observed.

REFERENCE

Gates, F. C. 1949. *Field Manual of Plant Ecology*. McGraw Hill Book Co., Inc., N. Y.

2.82 Energy Level of Standing Crop of the Sere

(Refer 1.82)

2.83 Succession Diagram

(Refer 1.83)

3 FRESHWATER ECOSYSTEM

3.00 HABITAT STUDY

3.01 Mapping of the Site

Procedure A (Contour map): Prepare the outline map of the water body as in exercise 1.01. Determine depth of the water at several points with the help of a boat and scaled sticks. These points should be at measured distances from marginal points. In the outline map put in the figures of depth and draw smoothed out lines. For mapping deep ponds wherein sticks cannot be placed, the boat is rowed slowly in a straight line, with the two ends marked on the shore. A prismatic compass with reference to two indicator points on the same side of the shore (particular tree, temple or rock), can also be used on the boat for taking bearings. A graduated string bearing a small hooked metal anchor is thrown ahead in the pond from the moving boat and the depth is read on the string as soon as it stands vertical. The anchor may then be dragged on the bottom for a metre or so in order to pull up submerged plants growing at that point. The process is repeated for a number of points in the pond.

(Refer 1.01 also)

This procedure can also be used for mapping any section of a large pond with open water on one side.

NOTE:—It is better to draw alongside a number of vertical sections of the pond based on the figures of depth. The level of water on a particular date should always be recorded.

The vegetation on the maps (prepared by any of the above methods) may be shown by using different symbols for species/communities in their respective boundaries.

3.02 Physical Characters of Soil

(Refer 1.02/A, B, C, D)

Procedure E: In ponds, lakes or rivers the soil and other particles gradually keep on settling at the bottom. This phenomenon is called silting. This can be measured by fixing some straight stick marked in centimeters vertically in the shallow regions of pond or lake. The initial reading of bottom is noted and at weekly intervals the rise in the bottom level is recorded to assess the thickness of silting, or the thickness of erosion,

Generally the silted soil is finer and darker in shade than the original bottom. Through soil auger, the column of bottom soil is sampled and thickness of silted layers (and weight of column of 1 cm$\times$1 cm) may be measured, if distinguishable. If a proper record of average annual thickness and weight of silt per unit column is made from the above observations, the age of the pond may be computed with some degree of precision.

3.03 Chemical Characters of Soil

(Refer 1.03)

3.04 Atmosphere

(Refer 1.04)

3.05 Biotic Factors

(Refer 1.05)

3.06 Water

A. TEMPERATURE

The temperature at the desired depths can be recorded in the following manner:

(1) A maximum-minimum thermometer is tied to a string to hang flat along with a weight attached to its base. Both the indicators are set to the present atmospheric temperature with the help of a magnet. The thermometer is quickly lowered to the desired depth and allowed to stand there for 10-15 minutes. The thermometer is then quickly pulled out and the temperatures recorded by both the indicators are read. It will be seen that one of the indicators remains at the point where it was set while the other one has moved. The indicator which has moved denotes the temperature of the water at the depth it was placed.

(2) To the bottom of a thermosflask a heavy weight is attached and to its mouth is fitted a rubber cork having two glass tubes to one of which is attached a long transparent polythene tube. The whole apparatus is attached to a long graduated string. The polythene tube is tightly closed with the help of a pinch-cock and the apparatus is lowered to the desired depth. The pinch-cock is opened so that the water is slowly drawn within the thermosflask. The water rises into the polythene tube up to the surface level showing that the thermosflask is completely filled with water. The thermosflask is then taken out and the temperature of the water in the thermosflask is determined with the help of a sensitive thermometer.

(3) Thermister is an apparatus which is supplied with a thermocouple attached

at the end of a long cable; the other end is attached to a milliameter calibrated directly in °C. The thermocouple is immersed to the desired depth and the temperature is read directly on the dial.

B. TURBIDITY

The turbidhity or clarity in pond or lake water depends upon the quantity of suspended microscopic soil particles and micro-organisms present. A circular disc painted in contrasting colour of black and white is fixed to a rope. The instrument called Secchi disc has a heavy weight attached beneath. The instrument is lowered down in water with the help of a scaled string till the colour contrast on the disc becomes indistinguishable. This depth is noted and used as an index of turbidity. The greater the depth the lower will be the turbidity value.

C. DIRECTION AND RATE OF FLOW

The direction and rate of flow of water in a river can be measured by throwing a piece of wood or large wood charcoal in different regions of a river. The time taken and distance travelled by the wood piece may be measured after a certain gap. Thus the speed of water (and direction through a magnetic needle) may be measured correctly.

The method to estimate volume of flow of water in rivers is described by Lemon (1962) and it is like this "In estimation of the volume of flow, the Embody formula is classic:

$$R = \frac{W \;\; DA \;\; L}{T}$$

where R=volume of flow in feet^3/second, W=average width of stream in feet, D= average depth in feet, A=constant (0.8 for a smooh bottom of mud, sand, hard pan, bedrock, or 0.9 for cobb!es and coarse gravel), and T=the average time (seconds) for a float to traverse the distance L in feet".

REFERENCE

Lemon, P. C. 1962. *Field and Laboratory Guide for Ecology*. Burgess Pub. Co., Minneapolis.

D. HARDNESS

REAGENTS

(i) Hardness buffer. Dissolve 81 g of ammonium chloride in 600 ml of distilled water, add 27.6 ml of 20% ammonium sulphide $(NH_4)_2S$ and make up to a volume of 1200 ml with concentrated ammonium hydroxide.

(ii) Hardness indicator. Dissolve 1 g of Enochrome Black T in 25 ml water and add 2 ml of normal sodium carbonate. Dilute to 100 ml with isopropyl alcohol, shake to mix and filter.

(iii) Hardness reagent. Dissolve 4 g of versene acid (ethylene diamine tetra-acetic acid), 1.7 g of sodium hydroxide and 0.1 g of magnesium chloride ($MgCl_2 . 6H_2O$) in water and dilute to 1 litre. Standardize against N/100 calcium chloride solution and adjust the strength of hardness reagent to N/20 by proper dilution.

Pipette 50 ml water sample in 200 ml conical flask and add with stirring 0.5 ml hardness buffer and 5 drops of indicator solution. Titrate against hardness reagent till blue colour first appears.

CALCULATION

ppm total hardness = volume of hardness reagent (as calcium carbonate) ml $\times$ 20

REFERENCE

Wilde, S. A., G. K. Voigt and J. G. Iyer. 1964. *Soil and Plant Analysis for Tree Culture*. Oxford Publishing House, Calcutta.

E. SULPHATE

Dissolve 50 g barium chloride in water and dilute to 500 ml.

Evaporate to dryness 250 ml of water sample acidified with a few drops of concentrated hydrochloric acid. Cool and moisten the residue with a few drops of concentrated hydrochloric acid and add 25 ml water. Heat to boiling and filter through Whatman No. 42 filter paper. Wash the basin and the filter paper thrice with 10 ml water each time. Heat the combined filtrate to boiling and add drop by drop, with constant stirring, 10 ml of 10% barium chloride solution. Allow to stand for 1 hour and then filter through Whatman No. 42 filter paper. Wash the precipitate several times with warm water. Ignite the filter paper with precipitate in a weighed crucible and find out the weight of residue.

CALCULATION

ppm sulphate = wt. of residue $\times$ 1644.

REFERENCE

Theroux, F. R., E. R. Eldridge and W. L. Mallmann. 1943. *Laboratory Manual for Chemical and Bacterial Analysis of Water and Sewage*. McGraw Hill Book Co. Inc., New York.

F. CHLORIDE:

REAGENTS

(i) Standard (0.05N) silver nitrate solution. Dissolve exactly 8.494 g of silver nitrate in water and dilute to 1 litre.

(ii) Potassium chromate indicator. Dissolve 5 g potassium chromate in 80 ml water. Add saturated silver nitrate with constant stirring until a permanent red precipitate is formed. Filter the solution and dilute the filtrate to 100 ml.

(iii) 0.1 N Sulphuric acid.

(iv) 0.1 N Sodium carbonate.

(v) Phenolphthalein indicator.

Pipette 50 ml of water sample into a basin and add a few drops of phenolphthalein. Adjust the pH of the solution to 8.2 (just colourless to phenolphthalein) by addition of sulphuric acid or sodium carbonate solution whichever necessary. Add 1 ml of chromate indicator and titrate against standard silver nitrate solution, till first permanent red colour appears. Run a blank determination.

CALCULATION

ppm chloride = $T - B$
(in terms of Cl)

Where T = Volume of sample titration, ml

B = Volume of blank titration, ml

This formula is applicable only when 50 ml water sample has been taken.

REFERENCE

Jackson, M. L. 1962. *Soil Chemical Analysis.* Asia Publishing House, Bombay.

3.10 COMMUNITY STRUCTURE

(Refer 1.10)

3.11 Frequency

Procedure A: Throw a 'Kanti' (or anchor) at any spot in the water body. Pull it out and record the plants attached to it. Repeat this procedure several times so as to cover all the parts of the water body. Arrange the data in the following table:

	Number of Times Kanti was thrown									
Species	1	2	3	4	5	6	7	8	9	10	Total
A	+	—	—	+ etc.							
B											
C											
D											
E											
F											
etc.											

Treat the 'Kanti' as quadrat and calculate the percentage frequency of any species through dividing the number of times it occurred by the total number of times the kanti was thrown and multiplying this with 100.

Procedure B : Lay the quadrat (1×1 m) a number of times, preferably randomly and prepare the list of species encountered each time as in quadrat method of grassland ecosystem. When the pond or lake is much deep a boat may be used for the purpose. (Treat the data as in exercise no. 1.11).

NOTE:—The estimation of submerged species may be difficult by this method. If the water is not too deep it is possible to lay submerged quadrats of smaller size and to pluck all the vegetation within the quadrats in order to make the list outside. In case the water body is too deep some dredge may be used (Welch 1948).

INTERPRETATION

After surveying a number of ponds/water bodies prepare a list of species putting those at top with maximum frequency and those at bottom with minimum. Is there any relation of frequency of different species with the type, physical features or age of the water body? Discuss.

REFERENCE

Welch, P. S. 1948. *Limnological Methods.* The Blackiston Co., Toronto.

NOTE:—Frequency of marsh species to be determined as in exercise 1.11.

3.12 Density

Density gives a good idea of the importance of the species and coupled with frequency it provides a good measure of phytosociologic and habitat behaviour of different species.

Lay the quadrat at a number of places as in frequency determination and count the individuals of each species each time. Arrange the data in the following table:

(Number of individuals per quadrat)

Species	Quadrat Nos.							Total
A	1	2	3	4	5	6	7	
B								
C								
etc.								

Calculate density as number of individuals of a species per quadrat studied. Specify the dimension of the quadrat.

If the water body is too deep and contains many submerged plants use the steel or plastic cylinder instead of quadrat. Lower this cylinder to the bottom of the water body and collect all the plants coming within the cylinder. Count the individual of each species. Repeat the procedure several times and calculate the density as above.

NOTE:—Sometimes it may be necessary to use both the methods jointly. Study the free floating and emergent species through quadrats and submerged ones through steel cylinders or dredges. The quadrat size should then correspond to the cylinder size.

INTERPRETATION

Discuss the changes in density of species in different ponds or water bodies.

NOTE:—For studies on phytoplankton nets of different sizes and shapes are used. Wisconsin nets (both small and large) are often used and when plankton alongwith respective water sample is required Juday plankton trap may be used. Closing nets of several types are also used for the same purpose. Details of all these instruments and their use are given in Welch (1948). But in almost all the nets nearly 40% of the planktons are lost. The following procedure can be adopted to estimate the phytoplanktons of various depths. Collect 2 litres of pond water from a desired depth and add 10 ml of concentrated formalin. Take one litre of this water into a pipette 1.5 m long and 1.5 cm in diameter. This is allowed to stand for eight days and then 1/3rd of it is drawn into another pipette and 1 ml of concentrated formalin is added to it. Allow it to stand and then draw 10 ml of liquid from it, add a drop of formalin. In about 50 minutes 98% plankters settle down.

Cut 4 strips of coverglass and fix the same on a slide with the help of either canadabalsam or araldite in such a way that a cavity to hold about 0.5 ml of liquid is formed. This cavity is calibrated with the help of micropipette. Take the concentrated sample in the cavity, put a coverglass and carefully estimate the number of plankters per unit volume of the pond water.

REFERENCE

Welch, P. S. 1948. *Limnological Methods*. The Blakiston Company, Toronto.

NOTE:—Density of marsh species is determined as in exercise 1.12.

3.16 Charting and Mapping of Vegetation

(Refer 1.16)

3.17 Intra- and Inter-Specific Relations

(Refer 1.17. The pots set up for competition study are submerged in water)

3.19 Profile Diagram

Extend a string across the traverse of the pond by means of a boat.

Pull out the plants growing at different depths with the aid of 'Kanti'. Measure the height of the plants encountered and make a diagram giving suitable scale to show the plant species, their respective heights and the depth of the pond at which they grow.

INTERPRETATION

Does the aquatic flora change with increase in the depth of the pond?

3.20 BEHAVIOUR OF COMPONENT SPECIES (AUTECOLOGY) OF THE ECOSYSTEM

(Refer 1.20)

3.21 Geography, Evolution and Systematics

(Refer 1.21)

3.22 Structural and Physiological Characters and their Variations

(Refer 1.22)

3.23 Population Characteristics and Ecotypic Differentiation

(Refer 1.23)

3.24 Cytogenetical and Breeding Behaviour

(Refer 1.24)

3.25 Reproductive Capacity, Dormancy, Etc.

(Refer 1.25)

The seeds of water plants may lose their viability on drying, hence care should be taken in the study of these seeds.

3.26 Vegetative Propagation

(Refer 1.26)

There are special vegetative reproductive bodies called 'turions'—they may be studied as seeds.

3.27 Seedling Mortality and Growth Rate (Productivity) at different Stages and Seasonal Periodicity

(Refer 1.27)

3.30 FLOW OF ENERGY

(Refer 1.30)

3.31 Radiations (Light, Heat transfer)

(Refer 1.31/A, B & C for abovewater measurements)

In water both the light intensity and quality change with the change in the water depth. In most of the cases this becomes the limiting factor for the growth of the plants under deep waters. At a certain depth the light intensity remains at such a level where the respiration and photosynthesis become equal and then the plants cannot grow, this is termed the compensation point.

Procedure D (Measurement of light intensity by 'Secchi-disc'):

APPARATUS

Secchi-disc consists of a white plate of 20 cm diameter divided into 4 equal parts and the alternate quarters are painted black. At the base of it in the centre a heavy weight is attached so that it remains suspended horizontally in water.

The Secchi-disc is gradually lowered down in water till the visibility of the white part is lost, and depth of the secchi-disc from the surface of the water is measured. Now the disc is gradually raised up and the depth at which it becomes visible again is noted down. The average of these two readings gives the depth of light penetration from where the reflected light is perceived.

OBSERVATIONS

Record your data in the following table:

S.No.	Depth at which the disc becomes invisible (in cm)	Depth at which disc becomes visible again (in cm)	average
1.			
2.			

NOTE:—4 or 5 readings must be taken at a time and this should be repeated at different hours of the day.

RESULTS

Express the results in terms of cm of depths.

REFERENCE

MacLean, R. C. and W. R. I. Cook. 1943. *Practical Field Ecology*. Mathew Book Co.

Procedure E: A platinum wire is gradually lowered in water till it loses its visibility and is raised again till it becomes visible. Both the depths are measured in mm and recorded in a table, average of the two gives the value to be expressed in Result.

OBSERVATIONS

As in Procedure D.

RESULTS

Express the results in terms of the depth in mm and turbidity corresponding to it.

T	D	T	D	T	D	T	D
7	1,095	20	264	70	138	180	62.4
8	971	22	391	75	130	200	57.4
9	873	24	361	80	122	250	49.1
10	794	26	336	85	116	300	43.2
11	729	28	314	90	110	350	38.8
12	674	30	296	95	105	400	35.4
13	627	35	257	100	100	500	30.9
14	587	40	228	110	93	600	27.7
15	551	45	205	120	86	800	23.4
16	520	50	187	130	81	1,000	20.9
17	493	55	171	140	76	1,500	17.1
18	468	60	158	150	72	2,000	14 8
19	446	65	147	160	68.7	3,000	12.1

T=Turbidity in parts per million, and D=the vanishing depth of the wire in millimetres.

REFERENCE

MacLean, R. C. and W. R. I. Cook. 1943. *Practical Field Ecology*. Mathew Book Co.

Procedure F: A light meter consisting of a photocell sealed in a water-tight chamber with glass cover is suspended in such a manner that the plane of the photo-cell remains parallel to the surface of water. The terminals are connected with a long lead to the milliameter calibrated in terms of lux.

The photo-cell is gradually lowered down and the light intensity is measured at desired depths, during various hours of the day.

OBSERVATIONS

Record the data in the following table:

S.No.	Time	Depth	Light intensity in lux	Remarks

RESULT

Express the results in lux at various depths.

NOTE:—Rooted vegetation can be correlated with the penetration of light.

3.32 Annual Increment due to Producers

(Refer 1.32 A, B & C for emergent and marsh species).

In case of the aquatic ecosystem the light intensity as well as the light quality happen to change drastically with the increase in the depth. As such care must be taken to find out the rate of photosynthesis at various depths.

Procedure D (Macrophytes) :

(1) Take out the total biomass by Ekman dredge.

(2) Measure the area for which the dredge samples.

(3) Wash the plant material thoroughly and dry the material in hot-air oven and determine the dry weight.

(4) Calculate the dry weight per unit area.

(5) The observations must be repeated occasionally and then the increase of dry weight per unit area within a specified period may be computed.

OBSERVATIONS

Tabulate your observations as follows :

Date	Dry weight in g per square meter	Increase in dry weight in g per square meter

Compute the total increase per year.

RESULT

Express your results as increase in dry weight per unit area per year.

REFERENCE

Welch, P. S. 1948. *Limnological Methods.* The Blakiston Company, Philadelphia.

NOTE:—The average values for 4 or 6 determinations from various places in a pond must be considered in the results.

Procedure F (Microphytes) :

(1) Take one or two litres of water and concentrate that by filteration.

(2) Take the above concentrate and centrifuge, discard the upper clear liquid and transfer the material to a platinum dish.

(3) Place the dish in hot air oven maintained at 60°C and dry it to constant weight (this usually requires 24 hours' drying).

(4) Transfer this to the furnace maintained at 600°C for 1 hour and then weigh again.

(5) The difference in the weights of 3 and 4 gives the dry weight of the plankton.

(6) This weight is also corrected for the dissolved organic matter in water, by running a similar set without plankton of the lake water and subtracting the same from the previous one.

(7) The data are collected from various depths and places in the pond or lake and then an average is calculated on per unit volume of the water.

(8) This is repeated at regular intervals and then annual increment is computed.

S.No.	Date	Dry weight in g per liter	Increase in dry weight between previous and present reading in g

RESULT

Express your results as increase in dry weight of the microproducers per unit volume per year. By this method the weight of Z 00 plankters is also included. The sample should be taken at such a time when Z 00 plankters are less in surface water.

REFERENCE

Welch, P. S. 1948. *Limnological Methods.* The Blakiston Company, Philadelphia.

Procedure G (Assimilation by macrophytes):

APPARATUS

Two glass bottles of about 10 cm diameter and 250-300 ml capacity. The upper and lower surfaces are flat. The bottle has a neck at its one end which traps the oxygen released during photosynthesis. However, if such a bottle is unavailable, ordinary glass bottle may be used.

REAGENTS

(i) Manganous sulphate. 480 g $MnS0_4 . 44H_2O$ is dissolved and the volume is made up to 1 litre.

(ii) Alkaline Iodine. 700 g KOH and 150 g KI in distilled water and volume made up to 1 litre.

(iii) Concentrated sulphuric acid.

(iv) Sodium thiosulphate. 6.2 g in freshly boiled distilled water and volume made up to 1 litre. 25 g salicylic acid is added as preservative.

Standardization of sodium thiosulphate solution : Dissolve 0.2454 g $K_2 Cr_2 O_7$ (dried at 130°C for 2 hrs. and cooled in desiccator) in distilled water; make up to 200 ml. Take 100 ml of this solution (N/40) in a beaker cooled in ice bath; add 1 ml alkaline-iodine solution drop by drop and then 1 ml conc. $H_2 SO_4$. If the smell of iodine is detectable then repeat the whole process as under such a circumstance the iodine is lost. Titrate with sodium thiosulphate solution until iodine colour begins to disappear; add starch solution as indicator and continue titration until blue colour just disappears. If 10 ml sodium thiosulphate is neutralized by 10 ml dichromate solution then it is standard. 200 ml of water is titrated (refer 5-7 as given below) against this solution and the amount of sodium thiosulphate used on being multiplied with 0.698 gives ml of O_2 per litre.

(1) Bottles are suspended at various depths with twigs of the plants found in that region and left for 1 hour.

(2) This process is repeated at various hours of the day and at regular intervals.

(3) The macrophytes are obtained (exercise 3.32/D) and the dry weight per unit area at each level of each species is determined.

(4) The amount of oxygen evolved and collected in the bottle is measured and volume corrected to NTP.

(5) To the water of the bottle (volume already known) add 1 ml of manganous sulphate solution and 1 ml alkaline-iodine solution well below the surface of the water.

(6) Stopper the bottle and mix it thoroughly.

(7) Add 1 ml. conc. $H_2 SO_4$ to this solution, mix well and allow to stand for 10 minutes.

(8) Take 200 ml of sample in a conical flask and titrate with N/40 sodium thiosulphate as described in standardization (any return of the blue colour after end point should be ignored).

(9) The estimation has to be made before and after the experiments to determine the increase in oxygen.

(10) A parallel experiment is run with dark bottle to determine the respiratory loss during this period.

(11) The dry weight of the twig is determined at the end of the experiment.

CALCULATION

Total amount of water in the bottle each time=a ml.

Volume of oxygen evolved in the gaseous form=b ml (corrected to NTP) in lighted bottle.

Volume of N/40 sodium thiosulphate required to neutralize 200 ml of water at the beginning of the experiment=c; in case of photosynthesis (lighted bottle) and in case of respiration (dark bottle)=d and e ml respectively after 1 hour each time.

Dry weight of the twig is X and Y g in light and dark bottles respectively.

Volume of O_2 at the beginning of the experiment in each bottle $= \frac{(c \times 0.698) \times a}{1000}$

Volume of O_2 at the end of the experiment in dark bottle $= \frac{(e \times 0.698)) \times a}{1000}$

Volume of O_2 at the end of the experiment in lighted bottle $= \frac{(d \times 0.698) \times a}{1000}$

Oxygen consumed by the twig in dark bottle

$$\left(\frac{(c \times 0.698)\, a - (e \times 0.698)\, a}{1000} \text{ ml}\right) \div Y = (R) \text{ per unit dry wt.}$$

Oxygen evolved in the lighted bottle

$$\left(\frac{(d \times 0.698)\, a - (c \times 0.698)\, a}{1000} \text{ ml} + b\right) \div X = (P\text{-}R) \text{ per unit dry wt.}$$

Therefore, total photosynthesis per unit dry weight

$= (P\text{-}R) + R$

OBSERVATIONS

Date	Time	Depth	Dry weight	Photo-synthesis	Total photo-synthesis per unit area	Total photo-synthesis
			1 2 3	1 2 3	1 2 3	
8.00 A.M.	d_1					
	d_2					
	d_3					
12.00 Noon	d_1					
	d_2					
	d_3					
4.00 P.M.	d_1					
	d_2					
	d_3					

Compute the total photosynthesis per unit area per year from these data.

RESULT

Express the results as volumes of O_2 evolved per unit area per year and from this compute the amount of energy fixed in terms of calories.

REFERENCES

Ikusima, Isac. 1965. Ecological studies on the productivity of aquatic plant communities. I. Measurement of photosynthetic activity. *Bot. Mag. Tokyo*. **78:** 202-211.

————1966. Ecological studies on the productivity of aquatic plant communities. II. Seasonal changes in standing crop productivity of *Vallisneria denseserrulata*. *Ibid.* **79:** 7-19.

Welch, P. S. 1948. *Limnological Methods*. The Blakiston Company, Philadelphia.

Procedure M (Assimilation by microphytes):

The procedure laid down in exercise 3.32/G has to be followed by taking water as sample containing microphytes in the bottle.

3.33 Food Chain and Rate of Removal of Produce by Consumers

(Refer 1.33)

The microconsumers may be estimated on per unit volume basis by using haemato-cytometer.

The population of the various fishes and other organisms may be determined by the method given in exercise 1.33/B & C of grassland ecosystem. Place some of these organisms in a water tank and find out their consumption per capita in terms of dry matter and with the help of this value calculate the total consumption of dry matter by the consumers.

To get an idea of the type of food consumed by these consumers the animal's food canal may be examined.

3.34 Energy Release by Decomposers

(Refer 1.34)

3.40 WATER CYCLE

(Refer 1.40)

3.41 Water Input (Precipitation)

(Refer 1.41)

3.42 Return of Water to Atmosphere

The magnitude of both the factors may be estimated by using hydrologic formulae as below:—

Evaporation = Precipitation + Run off + Amount of water present in the beginning of a period (WO) – [(Percolation + Transpiration + Amount of water present at the end of a period (Wt.)]

Similarly:

Transpiration=(Wo+P+RO)−(Wt+T+Percolation)

The procedure for the measurements of all the factors of the above mentioned equations have been described in exercise 1.42.

3.43 Loss of Water Due to Percolation

The equation describing hydrologic cycle may be used for indirect determination of the rate of percolation in a pond.

Precipitation+gain by surface run off +Amount of water present in the beginning of a period (WO)−[(Evaporation+Transpiration+Amount of water remaining after a period (Wt.)]

The procedure for the measurement of the factors of the hydrologic cycle has been described in exercise 1.40. The method to measure percolation through soil has been described in exercise 1.43.

3.44 Gain of Water Due to Run Off from Adjoining Area

In exercise 1.44 the method to determine the loss of water due to run off has been described. The same water gets accumulated in low-lying areas forming ponds. The same method employed in this case gives the gain of water due to run off from the adjoining area.

3.45 Conservation of Water in the Soil or Pond

For the measurements of various water relations of the soil refer the methods described in exercise 1.45/A-I.

Procedure J (Amount of water in pond):

(1) *Direct Measurement*: By surface mapping of a pond and by morphometry the surface area and the shape of the pond basin are respectively determined. This gives the total volume of the depression. The depth of water surface from the edge of the depression is determined and the volume of water is known by computation.

(2) *Indirect Measurement*: Isotope P^{32} of known strength is added in a pond. Its strength is then measured by a Geiger Counter. The degree of dilution is computed in terms of the volume of water.

3.46 Conservation of Water in the Biomass

(Refer 1.46)

3.50 CARBON CYCLE

(Refer 1.50)

3.51 Carbon Dioxide Absorption by the Biomass

(Refer 3.32/G & H)

3.52 Carbon Dioxide Loss from the Biomass

(Refer 3.32/G & H)

3.53 Carbon Dioxide Loss by Soil and Water

(Refer 1.53/A, B, & C for Soil)

Procedure D:

REAGENTS

(i) Phenolphthalein indicator. Dissolve 1 g in 200 ml of 50% alcohol and neutralise with 0.0227 N sodium hydroxide.

(ii) 0.0227 N sodium hydroxide solution.

Pour 100 ml of water sample in a Nesslers tube, taking care to avoid agitation. Add 10 ml of phenolphthalein indicator and titrate against standard (0.0227N) sodium hydroxide solution until faint permanent pink colour appears.

CALCULATIONS

ppm of carbon dioxide $=10 \times$ ml of 0.227N sodium hydroxide used in titration.

The periodic estimation of the CO_2 and total water mass will give the idea of loss and gain of CO_2 during that period.

REFERENCES

Welch, P. S. 1948. *Limnological methods*. The Blackiston Company, Philadelphia.

Welch, P. S. 1952. *Limnology*. MacGraw Hill Book Company, N.Y.

3.54 Storage of Carbon as Carbonate and Organic Matter in Soil and Water

(Refer 1.54)

3.60 NITROGEN CYCLE

3.61 Nitrogen Input by Nitrogen Fixing Organisms

For estimation of total nitrogen, ammonia and nitrate content in mud or bottom deposits, see the methods described in exercise 1.61/A-C. Water samples can also be analysed by the same methods, with slight modifications as described below.

TOTAL NITROGEN

Digest 200 ml of water sample in presence of a catalyst and sulphuric acid, as described in case of soils. Follow the rest of the procedure and calculate the amount of total nitrogen as follows:

$$\text{ppm of total nitrogen} = (T-B) \times \frac{1000}{200}$$

where T = sample titration, ml N/14 standard acid

B = blank titration, ml N/14 standard acid.

AMMONIA

Using 50 ml water sample proceed exactly as with soil extract. Determine the concentration of ammonia with reference to the calibration curve.

NITRATE

Evaporate 100 ml sample to dryness. Add phenoldisulphonic acid and proceed as in case of soils. Find out the concentration of nitrate in ppm from the calibration curve.

For estimations of the nitrogen input by nitrogen fixing organisms also refer exercise 1.61/D & E.

3.62 Nitrogen Loss by Denitrifying Organisms

(Refer 1.62)

3.63 Uptake and Storage of Nitrogen in Plants

(Refer 1.63)

3.70 CYCLING OF Ca, Mg, P, K

(Refer 1.70)

3.71 Uptake and Storage in Plants

(Refer 1.71)

3.72 Amounts Present in Soil and Water

Determination of calcium, magnesium, potassium, and phosphorus contents and pH of mud or botton deposits is made according to the procedures outlined in exercise 1.72. Water samples can be analysed by the same methods with slight modifications.

CALCIUM

Pipette 250 ml water sample into a silica basin, add few drops of concentrated hydrochloric acid and evaporate to dryness. Cool and add 10 ml of 6 N hydrochloric acid followed by 15 ml water and proceed as described for soils. Save the filtrate for magnesium estimation.

CALCULATIONS

ppm calcium =ml of 0.05 N potassium permanganate used in titration $\times 4$.

If very large amount of 0.05 N potassium permanganate is required for titration, use solution of higher normality, e.g. 0.125 N.

ppm calcium=ml of 0.125 N potassium permanganate used in titration $\times$ 10.

MAGNESIUM

Proceed with filtrate from calcium estimation as described for soils.

ppm magnesium = wt. of precipitate, g$\times$873.6.

POTASSIUM

Evaporate 250 ml water sample to dryness in a basin. Cool and dissolve the residue in exactly 15 ml of 0.17 N acetic acid and proceed as in case of soils.

CALCULATION

Calculate the amount of potassium in the aliquot as described for soils.

$$\text{ppm potassium} = \frac{\text{mg potassium in the aliquot}}{\text{volume of aliquot, ml}} \times 1000$$

PHOSPHORUS

Pipette an aliquot (10-25 ml) of water sample in a 50 ml volumetric flask, add 2-3 drops of dinitrophenol indicator and proceed exactly as in case of soils.

CALCULATION

Find out the concentration of phosphorus in the water sample directly from the calibration curve.

3.80 SUCCESSION

(Refer 1.80)

3.81 Recording the Data from the Sites showing Different Stages of Succession

(Refer 1.81)

3.82 Energy Level of Standing Crop of the Sere

(Refer 1.82)

In this case as the water stage ends and terrestrial begins, there is a sudden drop in energy content, which again rises later on (c.f. Lindeman, 1942).

REFERENCE

Lindeman, R. L. 1942. The trophic dynamic aspect of ecology. *Ecology* **23:** 399-418.

3.83 Succession Diagram

(Refer 1.83)

APPENDIX—A

List of Species Found in Grasslands of Banaras Hindu University

Name of the species	Germination	Flowering	Fruiting
Alloteropsis cimicina	June-July	July-Aug.	Aug.-Sept.
Alysicarpus longifolius	June-July	August	Sept.-Oct.
A. monilifer	June-July	Oct.-Dec.	Nov.-Mar.
Ammania baccifera	July	Aug.-Sept.	Oct.-Nov.
Amaranthus spinosus		Throughout the year	
Aneilema nudiflorum	June-July	August	Sept.-Oct.
Aristida adscencionis	June-July	August	Sept.-Oct.
Blumea sp.	October	Feb.-Apr.	Mar.-May
Bonnaya brachiata	June-July	July-Aug.	Aug.-Sept.
Boerhaavia diffusa	June-July	Throughout the year	
Bothriochloa pertusa	June-July	August	Sept.-Oct.
Cassia kleini	June-July	Aug.-Sept.	Sept.-Oct.
Cassia pumila			
C. tora	June-July	August	Sept.-Oct.
Caesulia axillaris			
Convolvulus pluricaulis	June-July	Apr.-May	May-June
C. arvensis	June-July	Oct.-Jan.	Feb.-Apr.
Corchorus acutangulus			
Crotalaria medicaginea	June-July	August	Sept.-Oct.
Croton sparciflorus	June-July	All the year round	
Cynodon dactylon			
Cyanotis axillaris			
Cyperus rotundus	Throughout the year	Aug.-Sept. Feb.-May	Aug.-Sept. Feb.-May
Dactyloctenium aegyptium	June-July	July-Aug.	Aug.-Oct.
Desmodium parvifolium			
D. triflorum	June-July	Oct.-Apr.	Nov.-May
Desmostachya bipinnata			
Dichanthium annulatum	June-July	Oct.-Apr.	Oct.-May
Digitaria granularis	June-July	Aug.-Sept.	Sept.-Oct.
D. marginata	June-July	Aug.-Sept., Oct.	Fruiting all the year round
D. royleana	June-July	July-Aug.	Aug.-Oct.
D. sanguinalis	June-July	July-Aug.	Aug.-Oct.
Echinochloa colonum	June-July	July-Aug.	Aug.-Oct.
Eclipta alba			
Eleusine indica	June-July	Aug.-Sept.	Oct.-Nov.
Eragrostis elongata			

Name of the species	Germination	Flowering	Fruiting
E. tenella	June-July	July-Aug.	Aug.-Sept.
E. viscosa			
Euphorbia hirta			
E. granulata			
E. hypericifolia	June-July	Almost throughout the year	
E. prostrata			
E. thymifolia			
Evolvulus alsinoides	June-July	Feb.-June	March-June
E. nummularis	June-July	Oct. onwards	
Fimbristylis diphylla	June-July	Aug., Sept., Oct.	
F. miliacea	June-July	Aug., Sept., Oct.	
F. schoenoides	June-July	Aug., Sept., Oct.	
Gnaphalium indicum	Sept.-Oct.	Oct.-Jan.	Feb.-Apr.
Heliotropium strigosum			
Heylandia latebrosa			
Imperata cylindrica			
Indigofera ennaephylla	June-July	Jan.-Mar.	Mar.-May
I. linifolia	June-July	Jan.-Mar.	Mar.-May
Kyllinga triceps	June-July	July-Aug.	Sept.-Oct.
Ludwigia parviflora			
Merremia emarginata			
Orthosiphon pallidus			
O. rubiculatus			
Oldenlandia corymbosa			
Panicum humile	June-July	July-Aug.	Sept-Oct.
P. trypheron			
P. psilopodium			
Paspalidium flavidum	June-July	July-Aug.	Sept.-Oct.
Paspalum scorbiculatum	June-July	Aug.-Sept.	Sept.-Oct.
Phyllanthus niruri	June-July	August	Aug.-Oct.
P. simplex	June-July	August	Aug.-Oct.
P. urinaria	June-July	August	Aug.-Oct.
Pulicaria crispa	July	Feb.-Mar.	Mar.-Apr.
Rhynchosia minima			
Rungia parviflora			

Name of the species	Germination	Flowering	Fruiting
Saccharum munja			
S. spontaneum			
Scoparia dulcis			
Setaria glauca	June-July	Aug.-Sept.	Sept.-Oct.
Sida veronicaefolia			
Sporobolus diander	June-July	July-Aug.	Aug.-Oct.
S. wallichii			
Tephrosia purpurea	March	Throughout the year	
Tribulus terrestris	March	Throughout the year	
Trichodesma indicum	October	Jan.-Apr.	May-June
Trianthema monogyna	July	Aug.-Sept.	October
Tridax procumbens			
Urochloa panicoides	June-July	July-Aug.	Aug.-Sept.
Urochloa reptans			
Vandellia crustacea	June-July	July-Aug.	Aug.-Sept.
Vernonia cinerea			
Volutarella divaricata	October	Nov.-Apr.	Mar-Apr..
Xanthium strumarium	In all the seasons		
Zornia diphylla	June-July	Aug.-Sept.	Sept.-Oct.

APPENDIX—B

List of Common Species found in Forests nearabout Varanasi

(Phenology given here is based on different published floras. The same should be checked for the area of study and correct time of phenological events like leaf fall, appearance of young foliage, time of flowering and fruiting, seed dispersal, etc. be noted).

Name of the species	Flowering	Fruiting	Remarks
	TREES		
Acacia arabica	June-Oct.	December	
A. catechu	July-Sept.	Cold season	
A. leucophloea	Aug.-Sep.	-do-	
A. intsia			
Adina cordifolia	June-July	Dec.-Mar.	
Aegle marmelos	May	Oct.-Nov.	Leafless Mar.-Apr.
Ailanthus excelsa	Apr.-May		Leafless in cold season
Albizzia lebbek	Apr.-May	Sept.-Mar.	
A. procera	May	Jan.-Sep.	
Anogeissus latifolia	Mar.-June	Nov.-Feb.	
A. pendula	Rainy season		Leafless in January
Antidesma diandrum	May-July	Aug.-Sep.	
Azadirachta indica	Cold and hot seasons	Ripe in July	Young leaves come out in beginning of hot season.
Bauhinia racemosa	Mar.-June	Nov.-Mar.	
B. tomentosa	Oct.-Nov.	Apr.-May	
Boswellia serrata			
Bridelia retusa	Apr.-Oct.	May-June	Leafless Mar.-Apr.
B. verrucosa			
Buchanania lanzan	Jan.-Mar.	Apr.-May	
Butea monosperma	Mar.	June-July	
B. parviflora			
Careya arborea	Mar.-Apr.	Ripe in July	
Casearia graveolens			
Cassia fistula	Apr.-May		
Cochlospermum religiosum	Feb.-Apr.		Fresh foliage in May
Cordia dichotoma	Mar.-Apr.	June.-Sept.	
C. rothu	Apr.-June	Cold season	
Dalbergia latifolia	Apr.-May	Dec.-Feb.	New leaves in Apr.-May

Name of the species	Flowering	Fruiting	Remarks
D. sissoo	Mar.-June	Nov.-Feb.	New foliage appears in February
Diospyros cordifolia	Mar.-Apr.	Ripe rainy season	
D. tomentosa	Apr.-May	Ripe Dec.	
Ehretia laevis	Jan.-July	Mar.-June	
Elaeodendron glaucum	Feb.-June	Sept.-Oct. to May	
Emblica officinalis	Mar.-May	Oct.-Feb.	
Erythrina suberosa	Mar.-May	May-June	
Feronia limonia	Hot season	Nov.-Jan.	
Flacourtia indica		Ripe Feb.-Apr.	
Ficus arnottiana		Ripe Feb.-Apr.	
F. bengalensis		Ripe Apr.-June	
F. glomerata		Ripe at different times throughout the year	
F. religiosa		Ripe Mar.-July	
F. tomentosa		Ripe Mar.-April	
Gardenia latifolia	Apr.-May	Next cold season	
G. turgida	Hot season	Rainy & cold season	
Garuga pinnata	Feb.-May	Rainy season	Leafless in dry seasons
Gmelina arborea	Mar.-Apr.	May-June	
Grewia tillaefolia	Mar.-May	Rainy season	Fresh foliage in April
Haplophragma adenophyllum			
Hesperethusa crenulata			
Holoptelia integrifolia	Feb.-Mar.	June-Aug.	
Hymenodictyon excelsum	June-July	October	
Kydia calycina	July-Jan.	Jan.-Feb.	
Lagerstroemia parviflora			
Lannea coromandelica			
L. grandis			
Madhuca indica	Mar.-Apr.	June-July	
Mallotus philippinensis	Nov.-Jan.	Ripe in Feb.-May	
Mangifera indica	Feb.-Mar.	Apr.-June	
Miliusa tomentosa			
M. velutina	Mar.-May	June-July	Leafless in early part of hot season.
Mimusops hexandra	Nov.-Dec.	Ripe in March	

Name of the species	Flowering	Fruiting	Remarks
Ougeinia oojeinensis	Mar.-May	June	Leafless in hot season
Phoenix acaulis	Cold season	Ripe May-June	
Pongamia pinnata	May-June	Next Apr.	
Prosopis juliflora	Feb.-Apr.	May-Aug.	
Pterocarpus marsupium	May-June	Dec.-Mar.	
Salmalia malabarica	Feb.-Mar.	Apr.-May	Leafless in Nov.-April
Sapium insigne	Cold season	March	
Schleichera oleosa	March		New leaves in March
Schrebera spp.			
Shorea robusta	March	Ripe in June	New leaves in March
Soymida febrifuga	Apr.-May	July-Aug.	Young foliage in April
Sterculia urens	Jan.	Apr.-May	Leafless in cold season
Stereospermum suaveolens	Apr.-May	Nov.-Dec.	
Streblus asper	Jan.-Mar.	Apr.-May	
Syzygium cumini	Mar.-May	May-Aug.	
Tamarindus indica			
Tectona grandis	June-Aug.	Ripe Nov.-Jun.	
Terminalia arjuna	Apr.-May	Cold season	
T.belerica	May	Next cold season	
T. chebula	Mar.-May	Dec.-Mar.	Leafless in Feb.
T. tomentosa	Apr.	Feb.-Apr.	Leafless in Mar.
	SHRUBS		
Adhatoda vasica	At different times throughout the year		
Balanites aegyptiaca	Mar.-May	Ripe in Dec.	New foliage in Mar.-May
Calotropis procera	At different places throughout the year		
Capparis zeylanica	Apr.-June	Nov.	
Carrisa congesta		Oct.-Jan.	
C. opaca			
Cassia sophera			
Clerodendron infortunatum	Dec.-Apr.	Rainy season	
Colebrookia oppositifolia	Dec.-Feb.	Mar.-Apr.	
Desmodium pulchellum	Rainy season	Rainy season	
Diospyros montana	Mar.-Apr.	Rainy season	

Name of the species	Flowering	Fruiting	Remarks
Euphorbia nivulia	March	Apr.-May	
Flemingia spp.			
Grewia flavescens			
G. hirsuta	Rainy season	Cold season	
Helicteres isora	Aug.-Dec.	Ripe Mar.	
Holarrhena antidysenterica	Mar.-June	Next Mar.	
Indigofera spp.			
Ixora sp.			
Murraya koenigi	Hot season	Rainy season	
Nyctanthes arbortristis	At different times throughout the year	Sep.-Jan. mainly	
Olax scandens	Cold season	Nov.-June	
Randia dumetorum	Mar.-June	Nov.-June	
R. uliginosa	Apr.-June	Dec.-Feb.	
Syzygium heyneanum	Mar.-Apr.	June	
Vangueria spinosa			
Vitex negundo	At different times throughout the year		
Woodfordia fruticosa	Oct.-Apr.		
Zizyphus glaberrima	Summer season	Ripe in cold season	
Z. nummularis	Mar.-June	Nov.-Jan.	

CLIMBERS

Name of the species	Flowering	Fruiting	Remarks
Abrus precatorius	Rainy season	January	
Acacia pennata	June-Aug.		
Ampelocissus latifolia	June-Aug.	Oct.	
Asparagus racemosus	Cold-season		
Butea superba	March		
Capparis spinosa	Apr.-June	Nov.	
Cissus repanda	Mar.-April	Apr.-May	New leaves in March
Cocculus hirsutus	Feb.-Mar.		
Combretum decandrum	Feb.-Mar.		

Name of the species	Flowering	Fruiting	Remarks
Ichnocarpus frutescens	Nov.-Dec.	Mar.-Apr.	
Jasminum multiflorum	Dec.-May	Dec.-May	
Milletia auriculata	Apr.-May	Next cold season	
Mimosa himalayana			
Piliostigma vahlii			
Smilax prolifera	Sep.-Oct.	Ripe in Dec.-Feb.	
Ventilago calyculata	Rainy season	Ripe in Feb.	
Zizyphus oenoplia	Sep.-Oct.	Feb.-Mar.	
	GRASSES		
Apluda mutica	Sept.-Nov.	Sept.-Nov.	
Aristida hystrix	Aug.-Sep.	Oct.-Nov.	
Bothriochloa pertusa	Aug.-Oct.	Oct.-Nov.	In protected & most places throughout the year
Cenchrus ciliaris			
Chrysopogon aciculatus	Aug.-Nov.	Aug.-Nov.	
C. fulvus			
Cynodon dactylon	At different times throughout the year		
Dendrocalamus hamiltonii			
D. strictus	Nov.		Deciduous in hot season
Desmostachya bipinnata			
Dichanthium annulatum	At different times throughout the year		Peaks in Oct., Dec., Feb., April.
Eleusine indica	At different times throughout the year		
Eragrostis gangetica }			
E. pilosa }	Aug.-Sept.	Sept.-Nov.	
E. tenella }			
E. uniloides }			
Erianthus munja			
Eulaliopsis binata			
Heteropogon contortus	Sept.-Dec.	Oct.-Jan.	
Imperata cylindrica	Apr.-May	Oct.-Dec.	
Iseilema laxum	Oct.-Jan.	Oct.-Jan.	

Name of the species	Flowering	Fruiting	Remarks
Phragmites karka	Sept.-Dec.		
Polypogon monspeliensis	Apr.		
Saccharum spontaneum	Aug.-Nov.	Aug.-Nov.	
Sorghum halapense	Oct.-Nov.	Oct.-Dec.	
Vetiveria zizanioides	Aug.-Jan.	Aug.-Jan.	
	HERBS		
Adiantum incisum			
A. philippense			
Actiniopteris radiata			
Bacopa monieri	At different times throughout the year		
Casia obtusifolia	Aug.	Aug.-Oct.	
C. tora	Aug.	Aug.-Oct.	
Cheilanthes tenuifolia			
Chlorophytum tuberosum			
Curculigo orchioides			
Cyperus niveus	Jul.-Sep.	Sep.-Oct.	
Evolvulus alsnioides	At different times throughout the year		
E. nummularis			
Fimbristylis junciformis	Jul.-Sep.	Sep.-Oct.	
Jussieua suffruticosa	Rainy season		
Ophioglossum fibrosum			
Polygonum glabrum			
Rauvolfia tetraphylla	Nov.-Dec.		
Scilla hyacinthina			
Selaginella bryopteris			
Sida spp.			
Solanum surattense	At different times throughout the year		
Tephrosia purpurea			
	PARASITES		
Cassytha filliformis			
Cuscuta reflexa	Nov.		
Dendrophthoe falcata			

APPENDIX—C

List of Species Found in Freshwater Ponds

FLOATING

Eichhornia crassipes
Hygroryza aristata
Lemna paucicostata
Neptunia oleracea
Pistia stratiotes
Spirodela polyrhiza
Wolffia arhiza
Azolla pinnata

SUBMERGED

Callitriche stagnalis
Ceratoplyllum demersum
Hydrilla verticillata
Hydrocharis dubia
Myriophyllum indicum
Myriophyllum spicatum
Najas graminea
N. minor
Nechamandra alternifolia
Ruppia maritima
Utricularia exoleta
Utricularia stellaris
U. flexuosa
Vallisneria spiralis
Zannichelia palustris

EMERGENT

Aponogeton crispum
A. natans
Blyxa octandra
Cyperus cephalotes
C. corymbosus
Limnophila aquatica
L. heterophylla
Monochoria vaginalis
Nelumbo nucifera
Nymphaea nochulli
Nymphaea stellata
Nymphoides cristatum
Nymphoides indicum
Ottelia alismoides
Paspalidium geminatum
Polygonum pulcherum
Potamogeton pectinatus
P. perfoliatus
Sagittaria sagittifolia
Tenagocharis latifolia
Trapa bispinosa

MARSH

Alternanthera sessilis
Asteracantha longifolia
Bergia capensis
Eriocaulon setaceum
Ipomoea reptans
Jussiaea repens
Nasturtium officinale
Polygonum glabrum
Scirpus articulatus

APPENDIX D

LIST OF EQUIPMENT

Abney level
Altimeter
Anemometer

Bags-polythene
Balance
Balance-Spring (1 Kg)
Barograph
Barometer
Basin silica
Basket wirenet
Battery storage
Bell jar
Blade
Boat
Boxes-light proof
Bulbs-incandescent
Bull lens
Burette-automatic
Burette-different sizes

Cage wire netting
Calippers
Camera
Centrifuge-moisture equivalent
Chamber celluloid, transparent
Cloth polythene
Cloth black
Colorimeter-Photoelectric
Conductivity meter
Conductivity cell
Core Sampler
Cork rubber
Crucibles silica
Crucibles Gooch

Dark Chamber
Dendrograph
Dendrometer
Dish Platinum
Distillation Unit
Drawing Board
Dry and Wet Bulb Thermometer

Flag

Flasks- Conical
Volumetric
Kjeldahl
Filter
Distillation
Separating
Florescent tubes
Flow meter
Flower pots
Forceps
Freezing mixture
Funnels
Funnel-Buchner
Furnace-Muffle

Gauze wire

Haematocytometer
Hoe
Hygrometer
Hypsometer
Incubator
India ink

Jar-W.M.

Kanti
Kjeldahl digestion and distillation outfit
Kjeldahl micro-
Knife

Luxmeter

Manure-farmyard
Maps—World and India
Maximum-minimum thermometer
Measuring tape
Measuring cylinder-250 ml
Mechanical shaker
Micrometer stage
Micrometer-ocular
Microscope—Compound
—ordinary
—binocular
Munsell soil colour chart

Nails
Needles
Nets-nylon holes 1 cm sq.

Oven

Pantograh
Paper—filter Whatman No. 1. & 42
—graph
—litmus
—pH
—cellophane red and blue
—ozalid
Pencil glass marking
Petridish
pH meter
Photometer
Pipette blood Collecting
Plant press
Platinum wire
Plate-glass
Plankton net
Point—centered-quarter
Point frame apparatus
Porcelain tile
Protractor
Pyrheliometer

Quadrat

Raingauze
Refrigerator
Ring infiltrometer
Ruler

Sand
Sap extractor
Secateur
Scissors
Secchi disc
Shaker rotary
Sieves
Spathula
Soil leaching assembly
Sponge
String
Stakes
Stadistet rod
Stands—burette
—funnel
Surveyor's needle

Test tubes
Thermometer—maximum-minimum
— ordinary
—Beckman
Tower-gas washing
Trays wooden (1 m × 1 m × 10 cm)
40 mesh wire net at the bottom
Trap-animal
Trough-glass
Trowel
Tube-calcium chloride
Tubes
Tubings— rubber
—glass
Turbidity meter

Vasculum

Wirenetting
Wash bottle
Weight box
Wide mouthed bottle
Wind vane

APPENDIX—E

LIST OF CHEMICALS

Acetone
Acid Sulphuric
  hydrochloric
  acetic
  phosphoric
  boric
  oxalic
  salicylic
  perchloric
Agar
Alcohol isopropyl
  ethyl
  isobutyl
Amonium acetate
  chloride
  hydroxide
  molybdate
  oxalate
  phosphate
  sulphide
  thiocyanate

Barium chloride
  hydroxide
  sulphate
Bromo-cresol green
Buffer solution

Calcium carbonate, hydroxide
Carbol erythrosin
Charcoal activated
Chloroform
Copper sulphate

Devarda's alloy
Dinitrophenal
Diphenyl amine

Eosin
Eriochrome Black T

Ferrous amonium sulphate
Freezing mixture

Gentian violet
Gibberellic acid
Glucose

Hydrogen per-oxide

Iodine

Magnesium chloride
  carbonate
  sulphate
Manganous sulphate
Mercuric oxide
  iodide
Methyl red

Peptone
Phenol
Phenolphthalein
Potassium dichromate
Potassium chromate/nitrate
Potassium hydroxide
  iodide
  hydrogen phosphate
  permanganate
  dihydrogen phosphate

Rose Bengal

Selenium dioxide
  powder
Silver nitrate
  sulphate
Sodium hydroxide
  thiosulphate
  carbonate
  tartarate
  chloride
  phosphate
  cobaltinitrate
Stannous chloride
Starch

Thiourea
Triphenyl tetrazolium Chloride
  (T. T. C.)

Universal indicator

Versene (Ethylene diamine tetra-
  acetic acid)

Xylol

Zinc-granulated

ERRATA

page	*for*	*read as*
178	2.16	2.17
179	2.17	2.18
180	2.18	2.19

SUBJECT INDEX